Contents

How to use this guide 2

Preface 3

Introduction 4

Part 1 Fire risk assessment 9

Step 1 Identify fire hazards 12

Step 2 Identify people at risk 14

Step 3 Evaluate, remove, reduce and protect from risk 15

Step 4 Record, plan, inform, instruct and train 36

Step 5 Review 42

Part 2 Further guidance on fire risk assessment and fire precautions 45

Section 1 Further guidance on fire risks and preventative measures 46

Section 2 Further guidance on fire-detection and warning systems 59

Section 3 Further guidance on firefighting equipment and facilities 63

Section 4 Further guidance on escape routes 67

Section 5 Further guidance on emergency escape lighting 104

Section 6 Further guidance on signs and notices 106

Section 7 Further guidance on recording, planning, informing, instructing and training 109

Section 8 Quality assurance of fire protection equipment and installation 117

Appendix A Example fire safety maintenance checklist 118

Appendix B Technical information on fire-resisting separation, fire doors and door fastenings 122

Appendix C Historic buildings 130

Appendix D Glossary 132

References 137

Further reading 141

Index 144

How to use the guide

This guide is divided into two parts:

- **Part 1** Explains what fire risk assessment is and how you might go about it. Fire risk assessment should be the foundation for all the fire precautions in your premises.
- **Part 2** Provides further guidance on fire precautions. The information is provided for you and others to dip into during your fire risk assessment or when you are reviewing your precautions.

The appendices provide example checklists, some detailed technical information on fire-resisting elements and advice on historic buildings.

This guide is one from a series of guides listed on the back cover.

The rest of this introduction explains how the law applies.

Technical terms are explained in the glossary and references to other publications listed at the end of the publication are identified by a superscript number in the text.

In this Guide reference is made to British Standards and standards provided by other bodies. The standards referred to are intended for guidance only and other standards could be used. Reference to any particular standard is not intended to confer a presumption of conformity with the requirements of the Regulatory Reform (Fire Safety) Order 2005 (the Order).[1]

The level of necessary safety (or service) must be dictated by the findings of your risk assessment so you may need to do more or less than that specified in any particular standard referred to. You must be prepared to show that what you have done complies with any requirements or prohibitions of the Order[1] irrespective of whether you have relied on a particular standard.

Preface

This guidance gives advice on how to avoid fires and how to ensure people's safety if a fire does start. Why should you read it? Because:

- Fire kills. In 2004 (England and Wales) fire and rescue services attended over 33,400 fires in non-domestic buildings. These fires killed 38 people and injured over 1,300.
- Fire costs money. The costs of a serious fire can be high and afterwards many businesses do not reopen. In 2004, the costs as a consequence of fire, including property damage, human casualties and lost business, were estimated at £2.5 billion.

This guide applies to England and Wales only. It does not set prescriptive standards, but provides recommendations and guidance for use when assessing the adequacy of fire precautions in premises providing residential care. Other fire risk assessment methods may be equally valid to comply with fire safety law. The guide also provides recommendations for the fire safety management of the premises.

Your existing fire safety arrangements may not be the same as the recommendations used in this guide but, as long as you can demonstrate that they meet an equivalent standard of fire safety, they are likely to be acceptable. If you decide that your existing arrangements are not satisfactory there may be other ways to comply with fire safety law. This means there is no obligation to adopt any particular solution in this guide if you prefer to meet the relevant requirement in some other way.

Where the building has been recently constructed or significantly altered, the fire detection and warning arrangements, escape routes and facilities for the fire and rescue service should have been designed, constructed and installed in accordance with current building regulations. In such cases, it is likely that these measures will be satisfactory as long as they are being properly maintained and no significant increase in risk has been introduced.

This guide should not be used to design fire safety in new buildings. Where alterations are proposed to existing premises, they may be subject to building regulations. However, it can be used to develop the fire safety strategy for the building.

Introduction

WHO SHOULD USE THIS GUIDE?

This guide is for all employers, managers, occupiers and owners of permanently staffed premises providing residential care where some or all of the residents might require assistance in the event of a fire; e.g. where residents may not be able to make their way to a place of total safety unaided. It tells you what you have to do to comply with fire safety law, helps you to carry out a fire risk assessment and identify the general fire precautions you need to have in place.

It applies to premises where the main use is the provision of residential care (where the primary purpose is to provide of personal and/or nursing care, not healthcare treatment). Typical residential care premises include those where care is provided for:

- the elderly or infirm;
- children and young persons;
- people with special needs such as those with learning difficulties or with mental or physical disabilities; and
- people with addictions.

This guide may also be suitable for individual residential care premises that are part of other multi-use complexes, although consultation with other people responsible will be necessary as part of an integrated risk assessment for the complex. The relevant parts of this guide can also be used as a basis for fire risk assessment in premises where care is provided on a non-residential basis, e.g. day care centres.

The guide is not intended for use in:

- sheltered accommodation, where no care is provided;
- premises where the primary use is healthcare treatment, e.g. hospitals (including private) and other healthcare premises; and
- single private dwellings where out-posted nursing care is provided.

Details of other guides in the series are given on the back cover.

The guide has been written to provide guidance for a responsible person, to help them to carry out a fire risk assessment in most residential care premises. If you read the guide and decide that you are unable to apply the guidance, then you should seek the expert advice of a competent person. Premises with very large numbers of residents (e.g. greater than 60), or with complicated layouts (e.g. a network of escape routes, or split levels), or those of greater than four storeys, or which form part of a multi-occupied complex, will probably need to be assessed by a competent person who has comprehensive training or experience in fire risk assessment. However this guide can be used for homes which are part of multi-occupied buildings to address fire safety issues within the individual occupancy.

This guide is intended to underpin national minimum standards.

It may also be useful for:
- employees;
- employee-elected representatives;
- trade union-appointed health and safety representatives;
- enforcing authorities; and
- all other people who have a role in ensuring fire safety in premises providing residential care.

If your premises are listed as of historic interest, also see Appendix C.

Fire safety is only one of many safety issues with which management must concern themselves to minimise the risk of injury or death to staff, residents or visitors. Unlike most of the other safety concerns, fire has the potential to injure or kill large numbers of people very quickly. This guidance is concerned only with fire safety but many of the measures discussed here will impact upon other safety issues, and vice versa; it is recognised that these various differing safety demands can sometimes affect one another and management should consult other interested agencies (e.g. the Health and Safety Executive (HSE)) where necessary to confirm that they are not contravening other legislation/guidance.

You can get advice about minimising fire losses from your insurer.

THE FIRE SAFETY ORDER

Previous general fire safety legislation

The Order[1] replaces previous fire safety legislation. Any fire certificate issued under the Fire Precautions Act 1971[2] will cease to have any effect. If a fire certificate has been issued in respect of your premises or the premises were built to recent building regulations, as long as you have made no material alterations and all the physical fire precautions have been properly maintained, then it is unlikely you will need to make any significant improvements to your existing physical fire protection arrangements to comply with the Order.[1] However, you must still carry out a fire risk assessment and keep it up to date to ensure that all the fire precautions in your premises remain current and adequate.

If you have previously carried out a fire risk assessment under the Fire Precautions (Workplace) Regulations 1997,[3] as amended 1999,[4] and this assessment has been regularly reviewed then all you will need to do is revise that assessment taking account of the wider scope of the Order[1] as described in this guide.

Introduction

The Order[1] applies in England and Wales. It covers 'general fire precautions' and other fire safety duties which are needed to protect 'relevant persons' in case of fire in and around most 'premises'. The Order[1] requires fire precautions to be put in place 'where necessary' and to the extent that it is reasonable and practicable in the circumstances of the case.

Responsibility for complying with the Order[1] rests with the 'responsible person'. In a workplace, this is the employer and any other person who may have control of any part of the premises, e.g. the occupier or owner. In all other premises the person or people in control of the premises will be responsible. If there is more than one responsible person in any type of premises (e.g. a multi-occupied complex), all must take all reasonable steps to co-operate and co-ordinate with each other.

If you are the responsible person you must carry out a fire risk assessment which must focus on the safety in case of fire of all 'relevant persons'. It should pay particular attention to those at special risk, such as disabled people (mobility impairment or learning disability), those who you know have special needs and children, and must include consideration of any dangerous substance liable to be on the premises. Your fire risk assessment will help you identify risks that can be removed or reduced and to decide the nature and extent of the general fire precautions you need to take.

If you or your organisation employ five or more people, your premises are licensed (including registered, e.g. with Commission for Social Care Inspection), or an alterations notice is in force, you must record the significant findings of the assessment. It is good practice to record your significant findings in any case.

There are some other fire safety duties you need to comply with:

- **You must** appoint one or more 'competent persons', depending on the size and use of your premises, to assist in undertaking any of the preventive and protective measures required by the Order[1] (you can nominate yourself for this purpose). A competent person is someone with enough training and experience or knowledge and other qualities to be able to implement these measures properly.

- **You must** provide your employees with comprehensible and relevant information on the risks to them identified by the fire risk assessment, about the measures you have taken to prevent fires, and how these measures will protect them if a fire breaks out.

- **You must** consult your employees (or their elected representatives) about nominating people to carry out particular roles in connection with fire safety and about proposals for improving the fire precautions.

- **You must**, before you employ a child, provide a parent with comprehensible and relevant information on the risks to that child identified by the risk assessment, the measures you have put in place to prevent/protect them from fire and inform any other responsible person of any risks to that child arising from their undertaking.

- **You must** inform non-employees, such as temporary or contract workers, of the relevant risks to them, and provide them with information about who are the nominated 'competent persons', and about the fire safety procedures for the premises.

- **You must** co-operate and co-ordinate with other responsible persons who also have premises in the building, inform them of any significant risks you find and how you will seek to reduce/control those risks which might affect the safety of their employees.

- **You must** provide the employer of any person from an outside organisation who is working in your premises (e.g. agency providing temporary staff) with clear and relevant information on the risks to those employees and the preventive and protective measures taken. You must also provide those employees with appropriate instructions and relevant information about the risks to them.

- If you are not the employer but have any control of premises which contain more than one workplace, **you are also responsible** for ensuring that the requirements of the Order[1] are complied with in those parts over which you have control.

- **You must** consider the presence of any dangerous substances and the risk this presents to relevant persons from fire.

- **You must** establish a suitable means of contacting the emergency services and provide them with any relevant information about dangerous substances.

- **You must** provide appropriate information, instruction and training to your employees, during their normal working hours, about the fire precautions in your workplace, when they start working for you, and from time to time throughout the period they work for you.

- **You must** ensure that the premises and any equipment provided in connection with firefighting, fire detection and warning, or emergency routes and exits are covered by a suitable system of maintenance and are maintained by a competent person in an efficient state, in efficient working order and in good repair.

- **Your employees must** co-operate with you to ensure the workplace is safe from fire and its effects, and must not do anything that will place themselves or other people at risk.

The above examples outline some of the main requirements of the Order.[1]
The rest of this guide will explain how you might meet these requirements.

Who enforces the Fire Safety Order?

The local fire and rescue authority (the fire and rescue service) will enforce the Order[1] in most premises. The exceptions are:

- Crown-occupied/owned premises where Crown fire inspectors will enforce;

- premises within armed forces establishments where the defence fire and rescue service will enforce;

- certain specialist premises including construction sites, ships (under repair or construction) and nuclear installations, where the HSE will enforce; and

- sports grounds and stands designated as needing a safety certificate by the local authority, where the local authority will enforce.

The enforcing authority will have the power to inspect your premises to check that you are complying with your duties under the Order.[1] They will look for evidence that you have carried out a suitable fire risk assessment and acted upon the significant findings of that assessment. If, as is likely, you are required to record the outcome of the assessment they will expect to see a copy.

If the enforcing authority is dissatisfied with the outcome of your fire risk assessment or the action you have taken, they may issue an enforcement notice that requires you to make certain improvements or, in extreme cases, a prohibition notice that restricts the use of all or part of your premises until improvements are made.

If your premises are considered by the enforcing authority to be high risk, they may issue an alterations notice that, amongst other things, requires you to inform them before you make any changes to your premises or the way they are used.

Failure to comply with any duty imposed by the Order[1] or any notice issued by the enforcing authority is an offence. You have a right of appeal to a magistrates court against any notice issued. Where you agree that there is a need for improvements to your fire precautions but disagree with the enforcing authority on the technical solution to be used (e.g. what type of fire alarm system is needed) you may agree to refer this for an independent determination.

If having read this guide you are in any doubt about how fire safety law applies to you, contact the fire safety office at your local fire and rescue service.

If your premises were in use before 2006, then they may have been subject to the Fire Precautions Act[2] and the Fire Precautions (Workplace) Regulations.[3,4] Where the layout (means of escape) and other fire precautions have been assessed by the fire and rescue service to satisfy the guidance that was then current, then it is likely that your premises already conform to many of the recommendations here, providing you have undertaken a fire risk assessment as required by the Fire Precautions (Workplace) Regulations.[3,4]

New buildings or significant building alterations should be designed to satisfy current building regulations which address fire precautions.[24] However, you will still need to carry out a fire risk assessment, or review your existing assessment (and act on your findings), to comply with the Order.[1]

Part 1 Fire risk assessment

MANAGING FIRE SAFETY

As the responsible person for premises providing residential care you should be fully aware of the need to manage your premises well to ensure the safety and well-being of your residents from fire at all times. Your staff must be trained to prevent or limit the risk of fire, recognise and neutralise potential fire hazards, and know how to respond to an emergency individually and collectively by actions and communications. You should keep your staff up to date on any issues that might cause a potential hazard or risk, and expect them to keep you informed when they spot problems. Where appropriate, you should also keep residents informed about hazards and risks and how to avoid them.

Good management of fire safety is essential to ensure that fires are unlikely to occur; that if they do occur they are likely to be controlled or contained quickly, effectively and safely; or that, if a fire does occur and grow, your staff are able to ensure that everyone in your premises is able to escape to safety easily and quickly, or remain in safety.

You therefore need to have robust and well-kept procedures to avoid fires occurring, to maintain the fire safety systems installed in your premises, to keep escape routes usable, to keep your staff up to date and well trained, and have emergency plans in place so that everyone (and in particular your staff, since they will have a critical role) know how to respond to a fire in your premises.

The risk assessment that you must carry out will help you ensure that your fire safety procedures, fire prevention measures, and fire precautions (plans, systems and equipment) are all in place and working properly, and the risk assessment should identify any issues that need attention. Further information on managing fire safety is available in Part 2 on page 45.

WHAT IS A FIRE RISK ASSESSMENT?

A fire risk assessment is an organised and methodical look at your premises, the activities carried on there and the likelihood that a fire could start and cause harm to those in and around the premises.

The aims of the fire risk assessment are:

- To identify the fire hazards.
- To reduce the risk of those hazards causing harm to as low as reasonably practicable.
- To decide what physical fire precautions and management arrangements are necessary to ensure the safety of people in your building if a fire does start.

The term 'where necessary' (see Glossary) is used in the Order,[1] therefore, when deciding what fire precautions and managment arrangements are necessary, you will need to take account this definition.

The terms 'hazard' and 'risk' are used throughout this guide and it is important that you have a clear understanding of how these should be used.

- **Hazard:** anything that has the potential to cause harm.
- **Risk:** the chance of that harm occurring.

If your organisation employs five or more people, or your premises are licensed or an alterations notice requiring it is in force, then the significant findings of the fire risk assessment, the actions to be taken as a result of the assessment and details of anyone especially at risk must be recorded. You will probably find it helpful to keep a record of the significant findings of your fire risk assessment even if you are not required to do so.

HOW DO YOU CARRY OUT A FIRE RISK ASSESSMENT?

A fire risk assessment will help you determine the chances of a fire starting and the dangers from fire that your premises present for the people who use them and any person in the immediate vicinity. The assessment method suggested in this guide shares the same approach as that used in general health and safety legislation and can be carried out either as part of a more general risk assessment or as a separate exercise. As you move through the steps there are checklists to help you.

Before you start your fire risk assessment, take time to prepare, and read through the rest of Part 1 of this guide.

Much of the information for your fire risk assessment will come from the knowledge your employees, colleagues and representatives have of the premises, as well as information given to you by people who have responsibility for other parts of the building. A tour of your premises will probably be needed to confirm, amend or add detail to your initial views.

It is important that you carry out your fire risk assessment in a practical and systematic way and that you allocate enough time to do a proper job. It must take the whole of your premises into account, including outdoor locations and any rooms and areas that are rarely used. If your premises are small you may be able to assess them as a whole. In larger premises you may find it helpful to divide them into rooms or a series of assessment areas using natural boundaries, e.g. areas such as kitchens or laundries, bedrooms, offices, stores, as well as corridors, stairways and external routes.

Under health and safety law (enforced by the HSE or the local authority) you are required to carry out a risk assessment in respect of any activities in your premises and to take or observe appropriate special, technical or organisational measures. If your health and safety risk assessment identifies that these activities are likely to involve the risk of fire or the spread of fire (for example in the kitchen or in a workshop) then you will need to take this into account during your fire risk assessment under the Order[1] and prioritise actions based on the level of risk.

You need to appoint one or more 'competent persons' (this could be you) to carry out any of the preventive and protective measures needed to comply with the Order.[1] This person could be an appropriately trained employee or, where appropriate, a third party.

Your fire risk assessment should demonstrate that, as far as is reasonable, you have considered the needs of all relevant people, including disabled people.

Figure 1 shows the five steps you need to take to carry out a fire risk assessment.

FIRE SAFETY RISK ASSESSMENT

1 Identify fire hazards
Identify:
Sources of ignition
Sources of fuel
Sources of oxygen

2 Identify people at risk
Identify:
People in and around the premises
People especially at risk

3 Evaluate, remove, reduce and protect from risk
Evaluate the risk of a fire occurring
Evaluate the risk to people from fire
Remove or reduce fire hazards
Remove or reduce the risks to people
- Detection and warning
- Fire-fighting
- Escape routes
- Lighting
- Signs and notices
- Maintenance

4 Record, plan, inform, instruct and train
Record significant finding and action taken
Prepare an emergency plan
Inform and instruct relevant people; co-operate and co-ordinate with others
Provide training

5 Review
Keep assessment under review
Revise where necessary

Remember to keep to your fire risk assessment under review.

Figure 1: The five steps of a fire risk assessment

STEP 1   IDENTIFY FIRE HAZARDS

For a fire to start, three things are needed:

- a source of ignition;
- fuel; and
- oxygen.

If any one of these is missing, a fire cannot start. Taking measures to avoid the three coming together will therefore reduce the chances of a fire occurring.

The remainder of this step will advise on how to identify potential ignition sources, the materials that might fuel a fire and the oxygen supplies that will help it burn.

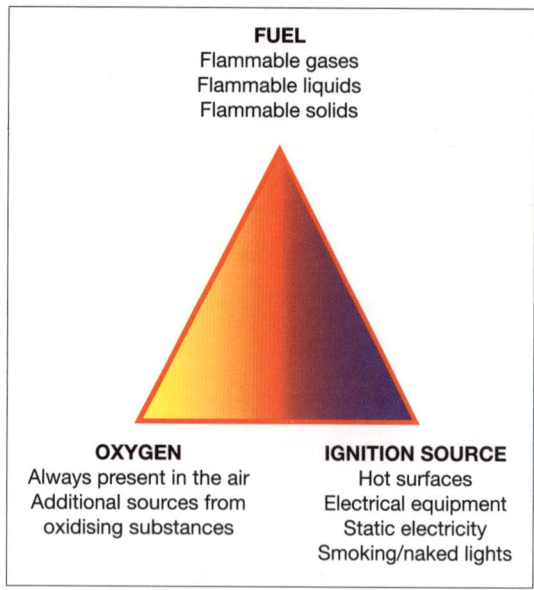

Figure 2: The fire triangle

1.1   Identify sources of ignition

You can identify the potential ignition sources in your premises by looking for possible sources of heat which could get hot enough to ignite material found in your premises. These sources could include:

- smoking materials, e.g. cigarettes, matches and lighters;
- naked flames, e.g. candles or gas or liquid-fuelled open-flame equipment;
- electrical, gas or oil-fired heaters (fixed or portable);
- cooking equipment;
- faulty or misused electrical equipment;
- lighting equipment;
- equipment owned or used by residents;
- hot surfaces and obstruction of equipment ventilation, e.g. photocopiers;
- hot processes, e.g. welding by contractors;
- arson, deliberate ignition, vandalism and so on.

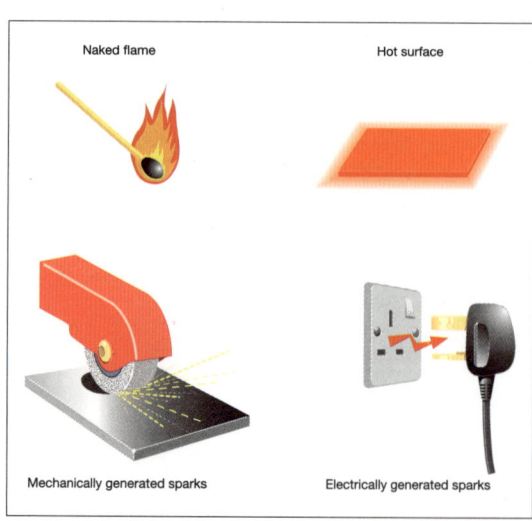

Figure 3: Sources of ignition

Indications of 'near-misses', such as scorch marks on furniture or fittings, discoloured charred electrical plugs and sockets, cigarette burns etc., can help you identify hazards which you may not otherwise notice.

1.2 Identify sources of fuel

Anything that burns is fuel for a fire. You need to look for the things that will burn reasonably easily and are in enough quantity to provide fuel for a fire or cause it to spread to another fuel source. Some of the most common 'fuels' found in premises providing residential care are:

- laundry supplies, such as bedding and towels, and medical supplies, such as disposable aprons;

- toiletries, aerosols;

- plastics and rubber (e.g. soft play or restraint areas), video tapes, polyurethane foam-filled furniture, foam-filled mats and polystyrene-based display materials;

- wood or wood-based furniture (permanent and temporary storage);

- textiles and soft furnishings, such as spare clothes and hanging curtains;

- private belongings, such as toys;

- seasonal and religious occasion decorations, such as Christmas decorations;

- items used in hobbies and crafts;

- flammable products, such as cleaning and decorating products, petrol, white spirit, methylated spirit, cooking oils, disposable cigarette lighters and photocopier chemicals;

- flammable gases such as liquefied petroleum gas (LPG), including aerosol canisters;

- paper products, packaging materials, stationery, advertising material, paper and books;

- waste products, particularly finely divided items such as shredded paper and wood shavings, off cuts, and dust; and

- waste storage, refuse containers and skips.

You should also consider the construction of your premises, and the materials used to line walls and ceilings, and how these might contribute to the spread of fire. You should check if the internal construction includes large areas of;

- hardboard, chipboard, block-board walls or ceilings;

- synthetic ceiling or wall coverings, such as polystyrene wall or ceiling tiles;

- flooring of polypropylene carpet or carpet tiles; or

- particular fixtures and fittings.

If these are present, and you are uncertain of the dangers they might pose, you should seek advice from a fire safety expert.

Further information is available in Part 2, Section 1.

1.3 Identify sources of oxygen

The main source of oxygen for a fire is in the air around us. In an enclosed building this is provided by the ventilation system in use. This generally falls into one of two categories: natural airflow through doors, windows and other openings; or mechanical air conditioning systems and air handling systems. In many buildings there will be a combination of systems, which will be capable of introducing/extracting air to and from the building.

Additional sources of oxygen can sometimes be found in materials used or stored at premises such as:

- some chemicals (oxidising materials), which can provide a fire with additional oxygen and so help it burn. These chemicals should be identified on their container (and Control of Substances Hazardous to Health data sheet, see Figure 4) by the manufacturer or supplier who can advise as to their safe use and storage; or

- oxygen supplies from cylinder storage and piped systems, e.g. medical oxygen, oxygen used by contractors (e.g. in welding processes).

Figure 4: Label on oxidising materials

Checklist

- Have you identified all potential ignition sources? ☐
- Have you identified all potential fuel sources? ☐
- Have you identified all potential sources of oxygen? ☐
- Have you made a note of your findings? ☐

STEP 2   IDENTIFY PEOPLE AT RISK

As part of your fire risk assessment, you need to identify those at risk if there is a fire. To do this you need to identify where you have service users residing and staff working (either at permanent workstations or at occasional locations around the premises) and to consider who else may be at risk, such as visitors, visiting contractors etc., and where these people are likely to be found.

You must consider all the people who use the premises but you should pay particular attention to people who may be especially at risk such as:

- employees who work alone, either regularly or at specific times, e.g. cleaners, security staff, maintenance staff, nursing or care staff (especially at night);

- people who are in isolated areas of your premises such as contractors working in roof spaces or plant rooms and maintenance staff;

- people who are unfamiliar with the premises, e.g. agency or temporary staff, guests, visitors (including visiting medical or social care staff), contractors;

- unaccompanied children and young persons;
- residents (clients and service users) who are unable to escape unaided (young children, babies, elderly, physically disabled (in particular people with mobility impairment), mentally disabled, people with vision or hearing impairment, those with some other sensory impairment, those whose ability to escape unassisted is impaired due to medication, or who who may be intoxicated);
- people who are not able to leave the premises quickly (but do not require assistance), e.g. residents or visitors who are elderly or with limited disabilities,* parents with children or children; and
- people who might panic or react adversely to the fire, the alarm or the excitement.

In evaluating the risk to people with disabilities you may need to discuss individual needs with each relevant person and record specific arrangements within the care plan for the individual. The risk assessment should take into account the resident's medical conditions, sensory awareness and mobility. In large residential care homes, especially those providing services for those with very severe conditions, you may also need to consult a professional access consultant or take advice from disability organisations*.

Further guidance on people with special needs is given in Part 2, Section 1.

Checklist
- Have you identified who is at risk? ☐
- Have you identified why they are at risk? ☐
- Have you made a note of your findings? ☐

STEP 3 EVALUATE, REMOVE, REDUCE AND PROTECT FROM RISK

The management of the premises and the way people use it will have an effect on your evaluation of risk. Management may be your responsibility alone or there may be others, such as the building owners or managing agents, who also have responsibilities. Some homes may be part of a complex and all those with some control must co-operate and consider the risk generated by others in the building.

To maintain a homely and non-institutional atmosphere, precautions should be introduced carefully, taking account of any possible adverse effects on the quality of service users' lives and the care they receive. For example, a self-closing door (which is a useful protection in the event of fire) is likely to be an inconvenience to older or disabled people or even cause an accident if care is not taken with the choice of door furniture (e.g. controlled free-swing self-closing device integrated with the automatic fire detection system) and its location.

*Visit the Disability Rights Commission website on www.drc-gb.org for more information.

3.1 Evaluate the risk of fire occurring

The chances of a fire starting will be low if your premises has few ignition sources and combustible materials are kept away from them.

In general, fires start in one of three ways:

- accidentally, such as when smoking materials are not properly extinguished or when bedside lights are knocked over;

- by act or omission, such as when electrical equipment is not properly maintained, or when waste is allowed to accumulate near to a heat source; and

- deliberately, such as an arson attack involving setting fire to external rubbish bins placed too close to the building.

Look critically at your premises and try to identify any accidents waiting to happen and any acts or omissions which might allow a fire to start. You should also look for any situation that may present an opportunity for an arsonist

Further guidance on evaluating the risk of a fire starting is given in Part 2, Section 1.

3.2 Evaluate the risk to people

In Step 2 you identified the people likely to be at risk should a fire start anywhere in the premises and earlier in Step 3 you identified the chances of a fire occurring. It is unlikely that you will have concluded that there is no chance of a fire starting anywhere in your premises so you now need to evaluate the actual risk to those people should a fire start and spread from the various locations that you have identified.

While determining the possible incidents, you should also consider the likelihood of any particular incident; but be aware that some very unlikely incidents can put many people at risk.

To evaluate the risk to people in your premises, you will need to understand the the way fire can spread. Fire is spread by three methods:

- convection;
- conduction; and
- radiation.

Convection

Fire spread by convection is the most dangerous and causes the largest number of injuries and deaths. When fires start in enclosed spaces such as buildings, the smoke rising from the fire gets trapped by the ceiling and then spreads in all directions to form an ever-deepening layer over the entire room space. The smoke will pass through any holes or gaps in the walls, ceiling and floor into other parts of the building. The heat from the fire gets trapped in the building and the temperature rises.

Conduction

Some materials, such as metal shutters and ducting, can absorb heat and transmit it to the next room, where it can set fire to combustible items that are in contact with the heated material.

Radiation

Radiation heats the air in the same way as an electric bar heater heats a room. Any material close to a fire will absorb the heat until the item starts to smoulder and then burn.

Smoke produced by a fire also contains toxic gases which are harmful to people. A fire in a building with modern fittings and materials generates smoke that is thick and black, obscures vision, causes great difficulty in breathing and can block the escape routes.

It is essential that the means of escape and other fire precautions are adequate to ensure that everyone can make their escape to a place of total safety before the fire and its effects can trap them in the building.

Figure 5: Smoke moving through a building

In evaluating this risk to people you will need to consider situations such as:

- fire starting on a lower floor affecting the escape routes for people on upper floors;

- fire developing in an unoccupied space that people have to pass by to escape from the building;

- fire or smoke spreading through a building via routes such as vertical shafts, service ducts, ventilation systems, poorly installed, poorly maintained or damaged walls, partitions and ceilings;

- fire and smoke spreading through a building due to poor installation of fire precautions, e.g. incorrectly installed fire doors (see Appendix B2 for more information on fire doors) or incorrectly installed services penetrating fire walls; and

- fire and smoke spreading through the building due to poorly maintained and damaged fire doors or fire doors being wedged open.

Further guidance on fire risks is given in Part 2, Section 1.

Where they suffer from limited mobility (or suffer claustrophobia), many residents in care homes may wish (or need) to keep their bedroom doors open, for ventilation or communications or comfort. Similarly, it can be of value to carers (and residents) to have doors open so that the carers can check the sleeping residents with a minimum of disturbance. A care home is not a prison; any fire safety systems should seek to not impair further the quality of life of the residents and the equipment provided and management procedures in place must be appropriate. In such cases, the use of 'hold-open' or 'free-swing' controlled door closing devices should be considered. See Appendix B3 for more information.

3.3 Remove or reduce the hazards that may cause a fire

Having identified the fire hazards in Step 1, you now need to remove those hazards if reasonably practicable to do so. If you cannot remove the hazards, you need to take reasonable steps to reduce them if you can. This is an essential part of fire risk assessment and as a priority this must take place before any other actions.

Ensure that any actions you take to remove or reduce fire hazards or risk are not substituted by other hazards or risks. For example, if you replace a flammable substance with a toxic or corrosive one, you must consider whether this might cause harm to people in other ways.

Remove or reduce sources of ignition

There are various ways that you can reduce the risk caused by potential sources of ignition, for example:

- Wherever possible replace a potential source by a safer alternative.

- Replace naked flame and radiant heaters with fixed convector heaters or a central heating system. Restrict the movement of and guard portable heating appliances.

- Operate a safe smoking policy in designated smoking areas and prohibit smoking elsewhere, ensuring sufficient ashtrays are provided and cleaned appropriately.

- Ensure electrical, mechanical and gas equipment is installed, used, maintained and protected in accordance with the manufacturer's instructions.

- Ensure all electrical fuses and circuit breakers are of the correct rating and suitable for the purpose and that electrical sockets are not overloaded.

- Check all areas where hot work (e.g. welding) has been carried out to ensure that no ignition has taken place or any smouldering materials remain that may cause of fire.

- Ensure that no one carrying out work on gas fittings which involves exposing pipes that contain or have contained flammable gas uses any source of ignition such as blow-lamps or hot-air guns.

- Operate a permit to work system for contractors who carry out hot work.

- Ensure that no one uses any source of ignition while searching for an escape of gas.

- Take precautions to avoid arson.

Remove or reduce sources of fuel

There are various ways that you can reduce the risks caused by materials and substances which burn, for example:

- Reduce the amount of combustible materials, such as paper products and plastics. Keep spare items in storerooms or storage areas where the residents or visitors are not allowed to go.

- Ensure combustible items, such as furniture, laundry and decorations are stored properly and are separate from potential ignition sources, such as boilers.

- Reduce quanitities of flammable liquids and gases in residents' areas to a minimum. Ensure spare supplies are kept to a minimum, and are stored properly in dedicated and locked storerooms or storage areas, preferably outside, where the residents and visitors are not allowed to go, and keep the minimum required for the operation of the home.

- Do not keep flammable solids, and flammable liquids and gases together.

- Make sure staff responsible for cleaning bedrooms are aware of potential fire hazards (e.g. storage, use and disposal of aerosols/newspapers) that may be brought into rooms by residents and visitors and left in a haphazard manner creating a fire risk. You should have a policy in place to deal with this constant hazard.

- Reduce or protect combustible displays, furnishings and foliage.

- Remove, cover or treat large areas of highly combustible wall and ceiling linings, e.g. polystyrene or carpet tiles, to reduce the rate of flame spread across the surface.

Figure 6: Storage of flammables

- Develop a formal system for the control of combustible waste by ensuring that waste materials and rubbish are not allowed to build up and are carefully stored until properly disposed of, particularly at the end of the day.

- Take action to avoid any parts of the premises and in particular storage areas being vulnerable to arson or vandalism.

- Check all areas where hot work (e.g. welding) has been carried out to ensure that no ignition has taken place and no smouldering or hot materials remain that may cause a fire later.

Further guidance on removing and reducing hazards is given in Part 2, Section 1.

Remove or reduce sources of oxygen

You can reduce the potential source of oxygen supplied to a fire by:

- Closing all doors, windows and other openings not required for ventilation or observation, particularly when staff levels are low.

- Shutting down ventilation systems which are not essential to the function of the premises.

- Not storing oxidising materials near or within any heat source or flammable materials.

- Controlling the use and storage of oxygen cylinders and/or piped oxygen, ensuring that they are not leaking, are not used to 'sweeten' the atmosphere, and that where they are located is adequately ventilated.

- Maintaining piped oxygen supplies in accordance with manufacturers' instructions.

3.4 Remove or reduce the risks to people

Having evaluated and addressed the risk of fire occurring and risk to people (preventive measures) it is unlikely that you will have concluded that no risk remains of fire starting and presenting a risk to people in your premises.

You now need to reduce the remaining fire risk to people to as low as reasonably practicable, by ensuring that adequate fire precautions are in place to warn people in the event of a fire and allow them to safely escape.

The rest of this step describes the fire protection measures you may wish to adopt to reduce the remaining fire risk to people (see Sections 3.4.1 to 3.4.6).

The level of fire protection you need to provide will depend on the level of risk that remains in the premises after you have removed or reduced the hazard and risks. Part 2, Section 4 on page 67 can help you decide the level of risk that you may still have.

Flexibility of fire protection measures

Flexibility will be required when applying this guidance; the level of fire protection should be proportional to the risk posed to the safety of the people in the premises. Therefore, the objective should be to reduce the remaining risk to a level as low as reasonably practicable. The higher the risk of fire and risk to life, the higher the standards of fire protection will need to be.

Your premises may not exactly fit the standards suggested in this guide and they may need to be applied in a flexible manner without compromising the lives of the occupants.

For example, if the travel distance is in excess of the norm for the level of risk you have determined (see Part 2, Table 2 on page 76), it may be necessary to do any one or a combination of the following to compensate:

- Provide earlier warning of fire using an enhanced automatic fire detection.
- Revise the layout to reduce travel distances.
- Reduce the fire risk by removing or reducing combustible materials and/or ignition sources.
- Control the number of people in the premises.
- Limit the area to trained staff only (no public or residents).
- Increase staff training and awareness.

Note: The above list is not exhaustive and is only used to illustrate some examples of trade-offs to provide safe premises.

If you decide to significantly vary away from the benchmarks in this guidance then you should seek expert advice before doing so.

3.4.1 Fire detection and warning systems

Virtually all premises that this guide applies to will need an electrial fire detection and warning system incorporating automatic fire detectors, sounders and manually operated call point (break-glass boxes).

In premises of limited size/occupation, e.g. ground and first floor with a small number of residents (up to four), a system of interconnected smoke alarms or point detectors incorporating interconnected manual call points and, where necessary separate sounders, may be accetable.

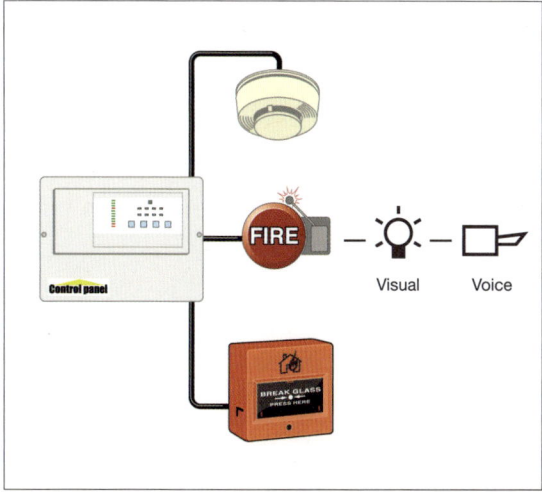

Figure 7: Fire detection and warning system

In larger premises with more residents, you will probably need an automatic fire detection and warning system with a control panel which is able to identify either zone or the specific location where the alarm has been raised.

In more complex premises, particularly those that accommodate a large number of residents over a number of floors, it is possible that a more sophisticated form of evacuation, e.g. phased evacuation, will be used. In these cases it will be necessary for the fire detection and warning systems to automatically trigger the action that people need to take (in accordance with the pre-determined fire procedure) e.g. by a two stage audible warning or voice alarm system.

You need to consider how residents, in particular non-ambulant residents, would or should respond to an alarm. Residents who require carer assistance to make their escape can do little except wait for rescue and the sound of the alarm could be distressing. The sound-level provided for residents then becomes of less significance. Communication procudures should be in place so that residents, who can hear the alarm but no respond, are notified as quickly as possible about what is happening.

It is of critical importance that staff know how to respond to an alarm and have well-rehearsed procedures in place.

Although the main risk will be to people when they are asleep you may still need to consider special arrangements for times when people are working alone, or are disabled or when your normal occupancy patterns are different, e.g. maintenance staff or other contractors working in the building.

False alarms from electrical fire detection and warning systems are a major problem (e.g. false activation of a detector due to poor system design) and result in many unwanted calls to the fire and rescue service every year. To help reduce the number of false alarms, the design and location of activation devices should be reviewed against the way the premises are currently used.

Further guidance on fire detection and warning systems is given in Part 2, Section 2.

Checklist

- Can the existing means of detection ensure a fire is discovered quickly enough for the alarm to be raised in time for all the occupants to escape to a place of safety? ☐
- Are the detectors of the right type and in the appropriate locations? ☐
- Can the means of warning be clearly heard and understood by everyone throughout the whole building when initiated from a single point? Are there provisions for people in locations where the alarm cannot be heard? ☐
- If the fire detection and warning system is electrically powered, does it have a back-up power supply? ☐

3.4.2 Firefighting equipment and facilities

Firefighting equipment can reduce the risk of a small fire, e.g. a fire in a waste paper bin, developing into a large one. The safe use of an appropriate fire extinguisher to control a fire in its early stages can also significantly reduce the risk to other people in the premises by allowing people to assist others who are at risk.

This equipment will need to comprise sufficient portable extinguishers that must be suitable for the risk.

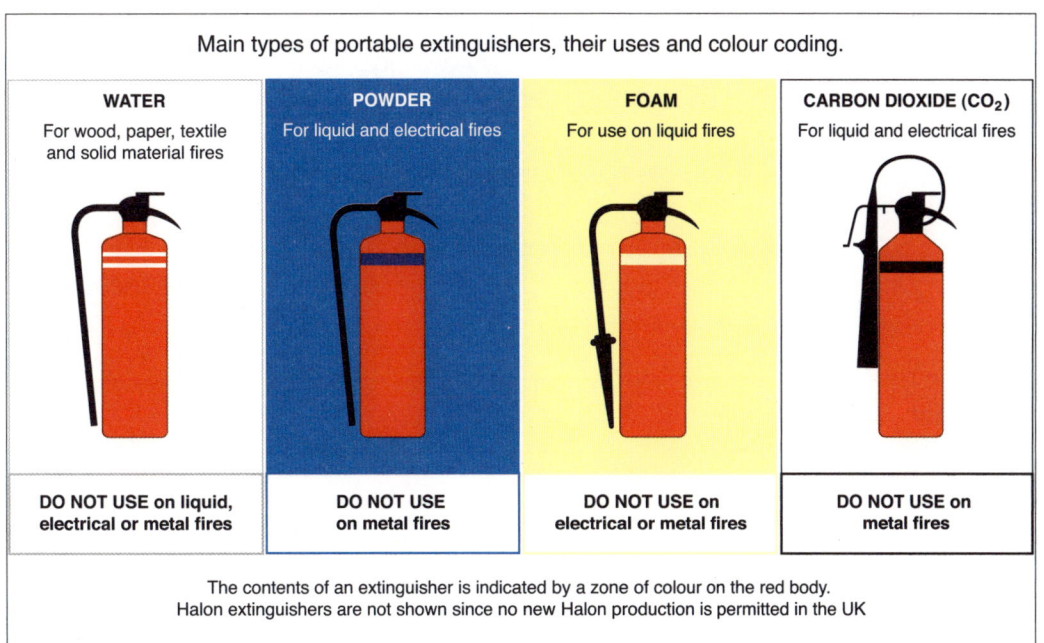

Figure 8: Types of fire extinguishers

In small premises, having one or two portable extinguishers of the appropriate type, readily available for use, may be all that is necessary. In larger, more complex premises, more portable extinguishers may be required and they should be sited in suitable locations such as on the escape routes at each floor level. It may also be necessary to indicate the location of extinguishers by suitable signs.

Some premises will also have permanently installed firefighting equipment such as hose reels, for use by trained staff or firefighters.

People with no training should not be expected to attempt to extinguish a fire. However, all staff should be familiar with the location and basic operating procedures for the equipment provided, in case they need to use it. If your fire strategy means that certain people, e.g. fire marshals, will be expected to take a more active role, then they should be provided with more comprehensive training.

In residential care homes the emphasis must be towards the safety of residents rather than fighting the fire; extinguishers should primarily be used to protect life and facilitate safe escape. They should otherwise only be used if they can be used safely and without risk of trapping the user.

Other fixed installations and facilities to assist firefighters, such as dry rising mains and access for fire engines, or automatically operated, fixed fire suppression systems such as sprinklers, may also have been provided.

Where these have been required by law, e.g. the Building Regulations or local Acts, such equipment and facilities must be maintained.

Similarly, if provided for other reasons, e.g. insurance, it is good practice to ensure that they are properly maintained.

In most cases it will be necessary to consult a competent service engineer. Keeping records of the maintenance carried out will help you demonstrate to the enforcing authority that you have complied with fire safety law.

Appendix A.1 provides a sample fire safety maintenance checklist you can use.

For more guidance on portable fire extinguishers, see Part 2, Section 3.1, for fixed firefighting installations, Part 2, Section 3.2 and other facilities (including those for firefighters), Part 2, Section 3.3.

Checklist

- Are the portable fire extinguishers or any fixed firefighting equipment provided suitable for controlling the risks identified?
- Are there enough extinguishers sited throughout the premises at appropriate locations?
- Are the right types of extinguishers located close to the fire hazards and can users get to them without exposing themselves to risk?
- Are the extinguishers visible or does their position need indicating?
- Have you taken steps to prevent the misuse of extinguishers?
- Do you regularly check equipment provided to help maintain the escape routes?
- Do you carry out daily checks to ensure that there is clear access for fire engines?
- Are those who test and maintain the equipment competent to do so?
- Do you have the necessary procedures in place to maintain any facilities that have been provided for the safety of people in the building (or for the use of firefighters, such as access for fire engines and firefighting lifts)?

3.4.3 Escape routes

Once a fire has started, been detected and a warning given, everyone in your premises should be able to escape to a place of total safety, either unaided or with assistance, but without the help of the fire and rescue service. However, some people with disabilities may need help from staff who will need to be designated for the purpose.

Escape routes should be designed to ensure, as far as possible, that any person confronted by fire anywhere in the building should be able to turn away from it and escape (or be taken) to a place of reasonable safety, e.g. a protected area or stairway. From there they will be able to go (or be taken) directly to a place of total safety away from the building.

Those who require special assistance (e.g. very young children, the elderly and infirm or people with disabilities) could be accommodated on the same level as the final exit from the premises to facilitate escape. Where they need assistance to evacuate, you should make sure that there are sufficient staff to ensure a speedy evacuation.

The level of fire protection that should be given to escape routes will vary depending on the level of risk of fire within the premises and other related factors. Generally, premises that are small, consisting of a single storey, will require fairly simple measures to protect the escape routes, but large or multi-storey buildings should have a more complex and inter-related system of fire precautions.

In all cases, escape routes should be designed to ensure, as far as possible, that any person confronted by fire anywhere in the building, should be able to turn away from it and escape (or be evacuated) to a place of reasonable safety, e.g. a protected area, corridor or stairway. From there, further escape should be possible either to another protected area or direct to a final exit.

When determining whether your premises have adequate escape routes, you need to consider a number of factors, including:

- the type, number and dependency of people using the premises;
- assisted means of escape;
- the evacuation strategy;
- escape time and travel distance;
- the age and construction of the premises;
- the number of escape routes and exits;
- the management of escape routes; and
- emergency evacuation of persons with mobility impairment.

The type, number and dependency of people using the premises

The people present in your premises will primarily be a mixture of staff, residents and visitors.

Staff can reasonably be expected to have an understanding of the layout of the premises, while visitors, particularly in larger premises, will be unlikely to have knowledge of alternative escape routes. Residents will probably know part of the premises and be familiar with some escape routes.

The number and dependency of people present will influence your assessment of the escape routes. You must ensure that your existing escape routes are sufficient and capable of safely evacuating all the people likely to use your premises at any time and particularly during times of high usage or at peak periods. If necessary you may need either to increase the capacity of the escape routes or restrict the number of people in the premises or consider other fire safety measures.

How quickly people can evacuate will depend on their level of reliance on staff, and it will therefore be helpful to consider the various dependencies:

- **Independent:** the mobility of residents is not impaired in any way and they are able to physically leave the premises without the assistance of staff or, if they experience some mobility impairment, they are able to leave with minimal assistance from another person.

- **Dependent:** all residents except those defined as independent or very high dependency. This category also includes those with mental health problems regardless of their independent mobility.
- **Very high dependency:** those residents whose care and / or condition creates a high dependency on staff and where immediate evacuation would prove potentially life threatening.

Assisted means of escape

People being cared for in residential care premises will vary considerably in terms of mobility and levels of awareness during a fire situation. There may be residents who exhibit severe mobility restriction, but will have a good awareness of the situation, being able to co-operate with staff. Others may exhibit normal mobility, but their level of awareness may be such that they present unpredictable behaviour, which may impede staff in an emergency. Children and residents with some forms of mental illness may become distressed by the alarm and the sudden activity.

The majority of care homes will provide care for individuals whose characteristics lie somewhere in between these examples. In almost all such cases, the evacuation of residents will rely entirely on the action of staff, moving them from the vicinity of the fire to an adjacent area of reasonable safety. Consequently it is essential that all staff, including agency and temporary staff, are fully trained in evacuation procedures for the premises.

The numbers of residents that may need to be moved in the event of a fire should be designed to be as few as possible. This may be achieved by establishing a number of protected areas within the premises. Restricting the number of residents (e.g. beds) within each protected area will be of benefit in an evacuation in terms of fewer residents requiring to be moved away from the fire.

You should ensure that staffing levels are sufficient and available at all material times to facilitate the movement of these numbers of residents to safety within the determined safe evacuation time (see below). Residents with severe difficulties for evacuation may be better situated at lower levels to aid the evacuation process.

Some residents in care homes may be confined to bed and subject to a range of procedures e.g. they may be catheterised, they may be receiving oxygen therapy, or they may be in receipt of automatic dosed medication. In such circumstances the attachment to medical equipment means that rapid evacuation may not be achieved within the required timescales. In such circumstances their bedroom is likely to be the most appropriate place of safety as part of delayed evacuation strategy (see below).

It is essential that your risk assessment is dynamic and takes full account of all difficulties that staff may encounter in moving people in an emergency. The problems that some residents may pose in an evacuation can change over short periods of time. Therefore it is important that staff play a part in the development of the emergency procedures. They will have a valuable contribution to make in terms of being aware of these changes, and be able to suggest any practical improvements to the evacuation procedure.

Evacuation strategy

In residential care premises, typical evacuation strategies are likely to involve one or more of the following arrangements:

- Single stage evacuation
- Progressive horizontal evacuation
- Delayed evacuation

Single stage evacuation

This strategy is appropriate for residents and others who fall predominantly into the 'independent' category, where it may reasonably be expected that all people in the building are able to (and will) evacuate immediately from the premises to a place of total safety without assistance.

Progressive horizontal evacuation

This strategy is likely to be necessary where the residents are dependent on staff to assist with their escape. It works on the principle of moving residents from an area affected by fire, through a fire resisting barrier to an adjoining fire protected area on the same level, where they can wait in a place of safety whilst the fire is dealt with, or await further evacuation down a protected route to total safety.

Delayed evacuation

Exceptionally, in some situations it may not be desirable or practical to evacuate some residents immediately (e.g. because of medical conditions or treatments). In these circumstances it may be appropriate to allow them to remain in their rooms whilst the fire is dealt with and the danger has passed, or to allow for the additional time necessary to prepare them for evacuation. In such circumstances, it will be necessary to provide enhanced levels of structural fire protection to the individual bedroom. However, where this strategy has been adopted, a suitable evacuation plan will still be required.

You should not depend upon the fire and rescue service to evacuate people; your escape strategy must be dependant only on the factors that are within your own control. However, you should consult with your enforcing authority when determining and planning an appropriate and effective evacuation strategy for your premises. Whichever system of evacuation you use must be supported by suitable management arrangements.

See Part 2 Section 4.1 for more information on evacuation strategies.

Escape time and travel distance

In the event of a fire, it is important to evacuate people as quickly as possible from the area or the building. Escape routes in a building should be designed so that people can escape quickly enough to ensure they are not placed in any danger from fire. The time available for escape will depend on a number of factors, including, how quickly the fire is detected and the alarm raised, the number of available escape routes, the volume and height of the space, the ventilation conditions, the speed of fire growth, the mobility of the residents and the availability of staff. For simplicity, travel distances in Part 2, Table 2 on page 76 take these factors into account.

At the design stage of a building, the length of escape routes is kept within certain guidelines so that in the event of a fire, the occupants of the premises can escape in the shortest time reasonably practicable.

The amount of escape time available will be closely linked to how quickly the staff and residents are made aware of a fire within the premises, combined with the distances to be travelled.

For single stage evacuation strategies, where it would be expected that all the occupants would evacuate immediately, either to a place of reasonable safety within the building e.g a protected stairway, or to a place of total safety outside the building, it is the time required to reach these points which is measured. However, where progressive horizontal evacuation is adopted, the escape time is related to how long it takes to move residents from the area involved in fire, to a place of reasonable safety in the adjoining protected area on the same level.

The time required to evacuate residents will be influenced by a number of factors, including:

- The degree of mobility of each resident to be moved;

- The level of awareness of each resident, and the level of co-operation that may be expected;

- The distance to be travelled to the adjoining protected area;

- The number of staff members available to move residents;

- The level of training given to staff in moving residents quickly in an emergency; and

- The need to disconnect any medical devices and to prepare residents for further evacuation.

The ideal way to determine the actual time required would be to conduct a timed simulation involving the actual movement of all of the residents. This would identify many simple problems that can be rectified before any emergency evacuation might be necessary. However, as this is not always practical, other methods may need to be developed for calculating escape time.

The age and construction of the premises

Older buildings may comprise different construction materials from newer buildings and may be in a poorer state of repair. The materials from which your premises are constructed and the quality of building work and state or repair could contribute to the speed with which any fire may spread, and potentially affect the escape routes the occupants will need to use. A fire starting in a building constructed mainly from combustible material will spread faster than one where fire-resisting construction materials have been used.

If you wish to construct internal partitions or walls in your premises, perhaps to divide up a recreation area or bedroom, you should ensure that any new partition or wall does not obstruct any escape routes or fire exits, extend travel distances or reduce the sound levels of the fire alarm system. Any walls that affect the means of escape should be constructed of appropriate material. Further technical information is provided in Appendix B.

Depending on the findings of your fire risk assessment, it may be necessary to protect the escape routes against fire and smoke by upgrading the construction of the floors, ceiling linings and walls to be fire-resisting. You should avoid having combustible wall and ceiling linings in your escape routes. For further guidance see Appendix B. You may need to seek advice from a competent person. Any structural alterations may require building regulation approval.

If your premises have been purpose built for the provision of residential care, and has been the subject of registration for this use in the past, it is likely that the level of structural protection present will be adequate. However, the number and size of protected areas that are available to staff for evacuation purposes may need to be reviewed if circumstances within the premises change significantly. An example may be an increase in the dependency of residents being cared for within the premises overall, e.g. a retirement home catering for the needs of fully ambulant residents in the past being used to accommodate largely non-ambulant residents.

If you have existing premises that are being used for residential care purposes for the first time, it is likely that upgrading of some structure will be required to support a policy of horizontal evacuation. For example, all floors should ideally have 60 minutes fire resistance, and a minimum of 30 minutes.

It is important that the correct balance is provided between adequate protection from fire, and maintaining a comfortable non-institutionalised environment.

The number of escape routes and exits

In general there should normally be at least two escape routes from all parts of the premises but a single escape route may be acceptable in some circumstances (e.g. part of your premises accommodating less than 60 people or where the travel distances are limited).

Where two escape routes are necessary and to further minimise the risk of people becoming trapped, you should ensure that the escape routes are completely independent of each other. This will prevent a fire affecting more than one escape route at the same time.

When evaluating escape routes, you may need to build in a safety factor by discounting the largest exit from your escape plan, then determine whether the remaining escape routes from a room, floor or building will be sufficient to evacuate all the occupants within a reasonable time. Escape routes that provide a single direction only may need additional fire precautions to be regarded as adequate.

Exit doors on escape routes and final exit doors should normally open in the direction of travel, and be quickly and easily openable without the need for a key. For premises where security is a major factor, (e.g. where elderly people may wander out of a building or children may get out of homes), you may have some form of security latches on final exit doors. In such cases you need to ensure that procedures are in place to ensure that all such doors are always available in an emergency, staff are properly trained and the procedures are included in your emergency plan.

Checks should be made to ensure final exits are wide enough to accommodate the number of people and/or wheelchair or evacuation chair users who may use the escape routes they serve.

Management of escape routes

It is essential that escape routes, and the means provided to ensure they are used safely, are managed and maintained to ensure that they remain usable and available at all times when the premises are occupied. Inform staff in training sessions about the escape routes within the premises.

Figure 9:
A blocked corridor
with incorrect signage

Corridors and stairways that form part of escape routes should be kept clear and hazard free at all times. Items that may be a source of fuel or pose an ignition risk should never be located on any corridor or stairway that will be used as an escape route. Have you made arrangements to ensure that all the escape routes within the building are properly maintained and available for use when required? See also Part 2, Section 4.

Emergency evacuation of persons with mobility impairment

The means of escape you provide must be suitable for the evacuation of everyone likely to be in your premises. This may require additional planning and allocation of staff roles – with appropriate training. Provisions for the emergency evacuation of disabled persons may include:

- stairways;

- evacuation lifts;

- firefighting lifts;

- horizontal evacuation;

- refuges; and

- ramps.

Use of these facilities will need to be linked to effective management arrangements as part of your emergency plan. The plan should not rely on fire and rescue service involvement for it to be effective.

Further guidance on escape routes is given in Part 2, Section 4.

Checklist

- Is your building constructed, particularly in the case of multi-storey buildings, so that, if there is a fire, heat and smoke will not spread uncontrolled through the building to the extent that people are unable to use the escape routes? ☐
- Are any holes or gaps in walls, ceilings and floors properly sealed, e.g. where services such as ventilation ducts and electrical cables pass through them? ☐
- Are there an adequate number of protected areas? ☐
- Can all the occupants escape to a place of total safety in a reasonable time? ☐
- Are the existing escape routes adequate for the numbers and type of people that may need to use them, e.g. residents and visitors, including disabled people? ☐
- Are the exits in the right place and do the escape routes lead as directly as possible to a place of total safety? ☐
- If there is a fire, could all available exits be affected or will at least one route from any part of the premises remain available? ☐
- Are the escape routes and final exits kept clear at all times? ☐
- Do the doors on escape routes open in the direction of escape? ☐
- Can all final exit doors be opened easily and immediately if there is an emergency? ☐
- Will everybody be able to safely use the escape routes from your premises? ☐
- Are your staff (and residents) aware of the importance of maintaining the safety of the escape routes, e.g. by ensuring that fire doors are not wedged open and that combustible materials are not stored within escape routes? ☐
- Are there any particular or unusual issues to consider? ☐

3.4.4 Emergency escape lighting

People in your premises must be able to find their way to a place of safety if there is a fire by using escape routes that have enough lighting.

In premises of limited size/occupation, e.g. ground and first floor, with a small number of residents (up to four) who do not rely upon staff for assistance with evacuation, and where the escape routes are simple and straightforward, borrowed lighting from a dependable source, e.g. from streetlamps where they illuminate the escape routes, may be acceptable.

In most care homes particularly those that are larger and more complex with more residents, a more comprehensive system of automatic emergency escape lighting should be in place to illuminate all the escape routes.

Where people have difficulty seeing conventional signs, a 'way-guidance' system may need to be considered.

Further guidance on emergency escape lighting is given in Part 2, Section 5.

Checklist

- Are all your escape routes covered by an acceptable form of emergency escape lighting? ☐
- Will there always be sufficient lighting to safely use escape routes? ☐
- Do you have back-up power supplies for your emergency escape lighting? ☐

3.4.5   Signs and notices

Signs

In care premises it is important to avoid an 'institutional' environment. However, signs must be used, where necessary, to help people identify escape routes, find firefighting equipment and emergency fire telephones. These signs are required under the Health and Safety (Safety Signs and Signals) Regulations 1996[5,6] and must comply with the provisions of those Regulations.

Other than in the smallest and simplest of premises where the exits are in regular use and familiar to staff, residents and visitors, a fire risk assessment that determines that no escape signs are required (because, for example, trained staff will always be available to help residents and visitors to escape routes), is unlikely to be acceptable to an enforcing authority.

For a sign to comply with these Regulations it must be in pictogram form (see Figure 10). The pictogram can be supplemented by text if this is considered necessary to make the sign more easily understood, but you must not have a safety sign that uses only text.

Where the locations of escape routes and firefighting equipment are readily apparent and the firefighting equipment is visible at all times, then signs are not necessary. In all other situations it is likely that the fire risk assessment will indicate that signs will be necessary.

Figure 10: Typical fire exit sign

Notices

Notices must be used, where necessary, to provide the following:

- instructions on how to use any fire safety equipment;
- the actions to be taken in the event of fire; and
- help for the fire and rescue service (e.g. location of sprinkler valves or electrical cut-off switches).

All signs and notices should be positioned so that they can be easily seen and understood. This may require essential signs to be duplicated at two levels (i.e. for ambulant and wheelchair residents).

Figure 11: Simple fire action notice

Further guidance on signs and notices is given in Part 2, Section 6.

Checklist

- Where necessary, are escape routes and exits, the locations of firefighting equipment, emergency telephones and other means of raising the alarm indicated by appropriate signs? ☐
- Have you provided notices such as those giving information on how to operate security devices on exit doors, those indicating doors enclosing fire hazards that must be kept shut and fire action notices for staff and other people? ☐
- Are you maintaining all the necessary signs and notices so that they continue to be correct, legible and understood? ☐
- Are you maintaining signs that you have provided for the information of the fire and rescue service, such as those indicating the location of water suppression stop valves and the storage of hazardous substances? ☐

3.4.6 Installation, testing and maintenance

New fire precautions should be installed by a competent person.

You must keep any existing equipment, devices or facilities that are provided in your premises for the safety of people, such as fire alarms, fire extinguishers, lighting, signs, fire exits and fire doors, in effective working order and maintain separating elements designed to and prevent fire and smoke entering escape routes.

You must ensure regular checks, periodic servicing and maintenance are carried out whatever the size of your premises and any defects are put right as quickly as possible.

You, or a person you have nominated, can carry out certain checks and routine maintenance work. Maintenance may need to be carried out by a competent service engineer.

Where contractors are used, third party certification is one method where a reasonable assurance of quality of work and competence can be achieved (see Part 2, Section 8).

The following are examples of checks and tests that should be carried out. The examples of testing and maintenance given are not intended to be prescriptive and other testing regimes may be appropriate.

Daily checks

Ensure that any security devices on fire exits do not impede escape (see Appendix B3 for more details), ensure that doors on escape routes swing freely and close fully and check exits and escape routes to ensure they are clear from obstructions and combustible materials and in a good state of repair. Check the fire alarm panel to ensure the system is active and fully operational. Where practicable, visually check that emergency lighting units are in good repair and apparently working. Check that all safety signs and notices are legible. Establish a routine for closing all compartmentation corridor fire doors at night within a pre-determined programme.

Weekly tests and checks

Test fire detection and warning systems and manually operated warning devices weekly following the manufacturer's or installer's instructions. Check that fire extinguishers and hose reels are correctly located and in apparent working order.

Monthly tests and checks

Test all emergency lighting systems to make sure they have enough charge and illumination according to the manufacturer's or supplier's instructions. This should be at an appropriate time when, following the test, they will not be immediately required.

Check that all fire doors in good working order and closing correctly and that the frames and seals are intact.

Six-monthly tests and checks

A competent person should test and maintain the emergency lighting and fire detection and warning system.

Annual tests and checks

All firefighting equipment, fire alarms and other installed systems should be tested and maintained by a competent person.

All structural fire protection and elements of fire compartmentation should be inspected and any remedial action carried out. Specific guidance on the maintenance of timber fire-resisting doors is given in Appendix B2.

Appendix A.1 provides an example of a fire safety maintenance checklist. You may find it useful to keep a log book of all maintenance and testing.

Further guidance on maintenance and testing on individual types of equipment and facilities can be found in the relevant section in Part 2.

Checklist

- Do you regularly check all fire doors and escape routes and associated lighting and signs? ☐
- Do you regularly check all your firefighting equipment? ☐
- Do you regularly check your fire detection and alarm equipment? ☐
- Are those who test and maintain the equipment competent to do so? ☐
- Do you keep a log book to record tests and maintenance? ☐

Step 3 Checklist

Evaluate, remove, reduce and protect from risks by:

- Evaluating the risk to people in your building if a fire starts. ☐
- Removing or reducing the hazards that might cause a fire. ☐

  Have you:

  – Removed or reduced sources of ignition? ☐
  – Removed or reduced sources of fuel? ☐
  – Removed or reduced sources of air or oxygen? ☐

  Have you removed or reduced the risks to people if a fire occurs by:

  – Considering the need for fire detection and for warning? ☐
  – Considering the need for firefighting equipment? ☐
  – Determining whether your escape routes are adequate? ☐
  – Determining whether your lighting and emergency lighting are adequate? ☐
  – Checking that you have adequate signs and notices? ☐
  – Regularly testing and maintaining safety equipment? ☐
  – Considering whether you need any other equipment or facilities? ☐

STEP 4 RECORD, PLAN, INFORM, INSTRUCT AND TRAIN

In Step 4 there are four further elements of the risk assessment you should focus on to address the management of fire safety in your premises. In smaller premises this could be done as part of the day-to-day management, however, as the premises or the organisation get larger it may be necessary for a formal structure and written policy to be developed. Further guidance on managing fire safety is given in Part 2, on page 45.

4.1 Record the significant findings and action taken

If you or your organisation employs five or more people, your premises are licensed, or an alterations notice requiring you to do so is in force, you must record the significant findings of your fire risk assessment and the actions you have taken.

Significant findings should include details of:

- The fire hazards you have identified (you don't need to include trivial things like a tin of solvent based glue).
- The actions you have taken or will take to remove or reduce the chance of a fire occurring (preventive measures).
- Persons who may be at risk, particularly those especially at risk.
- The actions you have taken or will take to reduce the risk to people from the spread of fire and smoke (protective measures).
- The actions people need to take in case of fire including details of any persons nominated to carry out a particular function (your emergency plan).
- The information, instruction and training you have identified that people need and how it will be given.

You may also wish to record discussions you have had with staff or staff representatives (including trade unions).

Even where you are not required to record the significant findings, it is good practice to do so.

In some simple premises providing residential care, record keeping may be no more than a few sheets of paper (possibly forming part of a health and safety folder), containing details of significant findings, any action taken, or proposed to be taken, and a copy of the emergency plan.

The record could take the form of a simple list which may be supported by a simple plan of the premises. (See Figure 12.)

In more complex premises it is best to keep a dedicated record including details of significant findings, any action taken, a copy of the emergency plan and maintenance of fire protection equipment and training. There is no one specified format for this. Further guidance is given in Part 2, Section 7.1.

You must be able to satisfy the enforcing authority, if called upon to do so, that you have carried out a suitable and sufficient fire risk assessment. Keeping records will help you do this and will also form the basis of your subsequent reviews. If you keep records, you do not need to record all the details, only those that are significant and the action you have taken.

It might be helpful to include a simple line drawing. This can also help you check your fire precautions as part of your ongoing review.

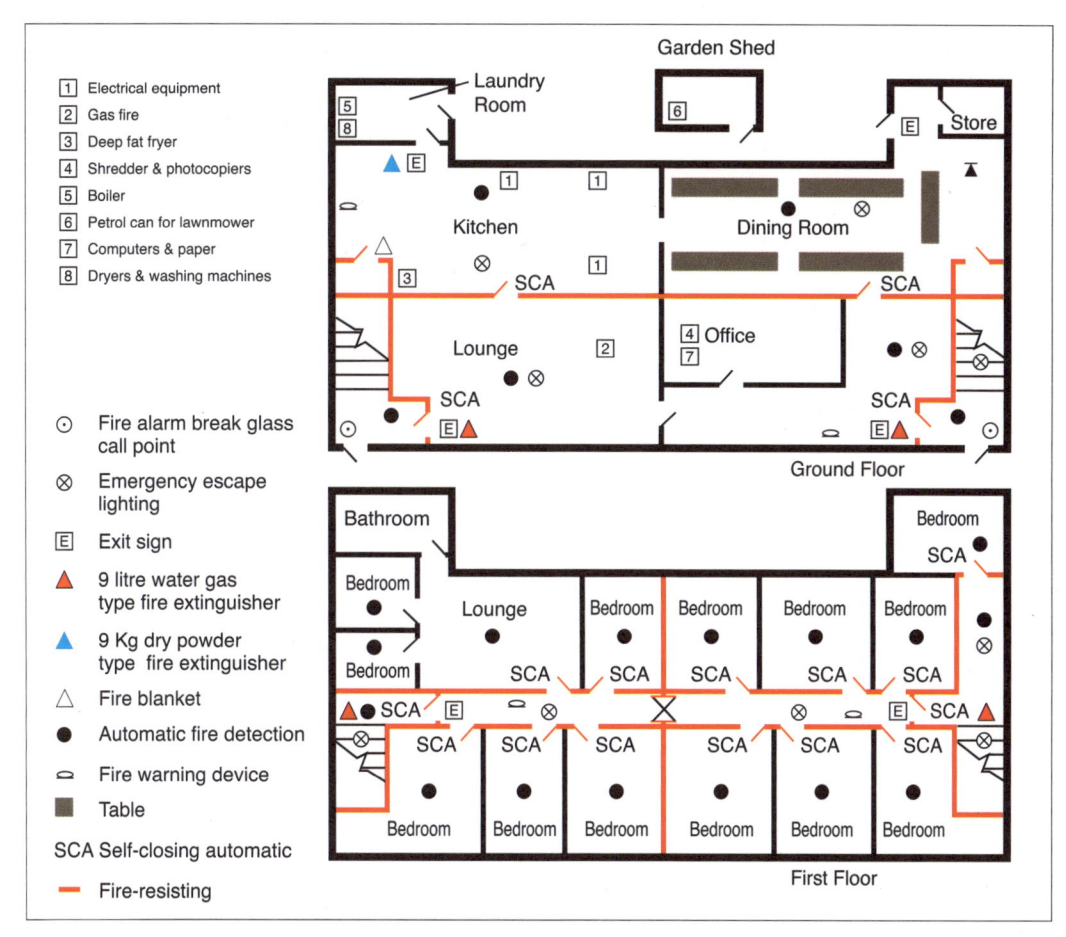

Figure 12: Example of a line drawing showing general fire safety precautions

The findings of your fire risk assessment will help you to develop your emergency plan, the instruction, information and training you need to provide, the co-operation and co-ordination arrangements you may need to have with other responsible people and the arrangements for maintenance and testing of the fire precautions. If you are required to record the significant findings of your fire risk assessment then these arrangements must also be recorded.

Records should also be kept of all alterations, tests, repairs and maintenance of fire safety systems, including passive systems such as fire doors, all 'near miss' events should be documented, and 'lessons learned'.

Further guidance about fire safety records with an example is given in Part 2, Section 7.1.

> **Checklist**
> - Have you recorded the significant findings of your assessment? ☐
> - Have you recorded what you have done to remove or reduce the risk? ☐
> - Are your records available for inspection by the enforcing authority? ☐

4.2 Emergency plans

You need to have an emergency plan for dealing with any fire situation. The purpose of an emergency plan is to ensure that all the people in your premises know what to do if there is a fire and that the premises can be safely evacuated.

If you or your organisation employ five or more people, or your premises are licensed or an alterations notice requiring it is in force, then details of your emergency plan must be recorded. Even if it is not required, it is good practice to keep a record.

Your emergency plan should be based on the outcome of your fire risk assessment and be available for your employees, their representatives (where appointed), residents (if they request it) and the enforcing authority.

In small premises providing residential care the emergency plan may be no more than a fire action notice.

In most premises providing residential care, the emergency plan will need to be more detailed.

Further guidance on emergency plans is given in Part 2, Section 7.2.

> **Checklist**
> - Do you have an emergency plan and, where necessary, have you recorded the details? ☐
> - Does your plan take account of other emergency plans applicable to the same building? ☐
> - Is the plan readily available for staff to read? ☐
> - Is the emergency plan available to the enforcing authority? ☐

4.3 Inform, instruct, co-operate and co-ordinate

You must give clear and relevant information and appropriate instructions to your staff and the employers of other people working in your premises, such as contractors, about how to prevent fires and what they should do if there is a fire. In some care homes you may also want to give information to residents and regular visitors.

If you intend to employ a child, you must inform the parents of the significant risks you have identified and the precautions you have taken. You must also co-operate and co-ordinate with other responsible people who use or are connected to any part of the premises. It is unlikely that your emergency plan will work without this.

Information and instruction

All staff should be given information and instruction as soon as possible after they are appointed and regularly after that. Make sure you include staff who work outside normal working hours, such as contract cleaners or maintenance staff.

All other relevant persons should be given information about the fire safety arrangement as soon as possible, e.g. residents when they take up residency. Information should be provided for the public (visitors) e.g. fire action notices.

The information and instructions you give must be in a form that can be used and understood. They should take account of those with disabilities such as hearing or sight impairment, those with learning difficulties and those who do not use English as their first language.

The information and instruction you give should be based on your emergency plan and must include:

- the significant findings from your fire risk assessment;
- the measures that you have put in place to reduce the risk;
- what staff should do if there is a fire;
- the identity of people you have nominated with responsibilities for fire safety;
- the importance of (closed) doors; and
- any special arrangements for serious and imminent danger to persons from fire.

In most care homes, written instructions should be provided to your staff who have been nominated to carry out a designated safety task, such as calling the fire and rescue service or checking that exit doors are available for use at the start of each shift.

This should include showing staff the fire-protection arrangements, including the designated escape routes, the location and operation of the fire-warning system and any other fire-safety equipment provided, such as fire extinguishers, and how to care for and evacuate residents. Fire action notices can complement this information and where used, should be posted in prominent locations.

Further guidance on information and instruction to staff, and on working with dangerous substances is given in Part 2, Section 7.3.

Co-operation and co-ordination

In some premises (e.g. owner-occupied premises) you may be solely responsible. However, in buildings owned by someone else, or where there is more than one occupier, and others are responsible for different parts of the complex, it is important that you liaise with them and inform them of any significant risks that you have identified. By liaising you can co-ordinate your resources to ensure that your actions and working practices do not place others at risk if there is a fire, and a co-ordinated emergency plan operates effectively.

Employees also have a responsibility to co-operate with their employer so far as it is necessary to help the employer comply with any legal duty.

Further guidance on co-operation and co-ordination is given in Part 2, Section 7.3.

> **Checklist**
> - Have you told your staff about the emergency plan, including the identity of people you have nominated to do a particular task? ☐
> - Have you told residents about the emergency plan? ☐
> - Have you given staff information about any dangerous substances? ☐
> - Do you have arrangements for informing temporary or agency staff? ☐
> - Do you have arrangements for informing other employers whose staff are guest workers in your premises, such as maintenance contractors and cleaners? ☐
> - Have you co-ordinated your fire safety arrangements with other responsible people and with any contractors in the complex? ☐
> - Have you recorded details of any information or instructions you have given and the details of any arrangements for co-operation and co-ordination with others? ☐

4.4 Fire safety training

You must provide adequate fire safety training for your staff. The type of training should be based on the particular features of your premises including any ancillary accommodation and should:

- take account of the findings of the fire risk assessment;
- explain your emergency procedures;
- take account of the work activity and explain the duties and responsibilities of staff;
- take place during their individual working hours and be repeated periodically where appropriate;
- be easily understandable by your staff and other people who may be present;
- be provided to new, agency and temporary staff prior to starting their duties;
- address the roles of staff and others (guests, residents); and
- be tested by fire drills.

In simple premises this may be no more than showing new staff the fire exits and giving basic training on what to do if there is a fire. In larger premises, with a high staff turnover and many shift patterns, the organisation of fire safety training will need to be planned.

Your training should include the following:

- the importance of keeping fire-doors closed (or closing them) to prevent the spread of fire, heat and smoke;

- what to do on discovering a fire;

- how to raise the alarm and what happens then;

- what to do upon hearing the fire alarm;

- when to adopt, and the procedures for, a 'delayed evacuation' response;

- the procedures for alerting other staff, residents and visitors including, where appropriate, directing them to exits;

- the arrangements for calling the fire and rescue service;

- the identification and use of protected areas for horizontal evacuation;

- the evacuation procedures for everyone in your premises to reach an assembly point at a place of total safety, in particular the role of residents;

- the evacuation procedures for residents who require assisted escape, to reach an assembly point at a safe place;

- the location and, when appropriate, the use of firefighting equipment;

- the location of escape routes, especially those not in regular use;

- how to open all emergency exit doors;

- where appropriate, how to stop machines, appliances and processes and isolate power supplies in the event of a fire;

- the reason for not using lifts (except those specifically installed or nominated, following a suitable fire risk assessment, for the evacuation of people with a disability);

- the safe use of and risks from storing or working with highly flammable and explosive substances and bottled or piped oxygen;

- the importance of general fire safety, which includes good housekeeping; and

- fire drills, with and without residents' involvement.

All the staff identified in your emergency plan that have a supervisory role if there is a fire (e.g. heads of department, fire marshals or wardens and, in larger premises providing residential care, fire parties or teams), should be given details of your fire risk assessment and receive additional training, and be aware of the importance of staff roles and staffing ratios.

Further guidance on fire safety training and examples of how to carry out a fire drill is given in Part 2, Section 7.4.

> **Checklist**
> - Have your staff received any fire safety training? ☐
> - Have you carried out a fire drill recently? ☐
> - Are staff aware of specific tasks if there is a fire? ☐
> - Are residents aware of specific actions if there is a fire? ☐
> - Are you maintaining a record of training sessions? ☐
> - If you use or store hazardous substances have your staff received appropriate training? ☐

STEP 5   REVIEW

You should constantly monitor what you are doing to implement the fire risk assessment, to assess how effectively the risk is being controlled.

If you have any reason to suspect that your fire risk assessment is no longer valid or there has been a significant change in your premises that has affected your fire precautions, you will need to review your assessment and if necessary revise it. Reasons for review could include:

- changes to work activities or the way that you organise them, including the introduction of new equipment;
- alterations to the building, including the internal layout;
- substantial changes to furniture and fixings;
- the introduction, change of use or increase in the storage of hazardous substances;
- the failure of fire precautions, e.g. fire-detection and alarm systems, life safety sprinklers or ventilation systems;
- significant problems reported by staff or residents;
- a significant increase in the number of people present;
- the presence of people with some different or specific form of disability; and
- changes in staff ratios.

You should consider the potential risk of any significant change before it is introduced. It is usually more effective to minimise a risk by, for example, ensuring adequate, appropriate storage space for an item before introducing it to your premises.

Do not amend your assessment for every trivial change, but if a change introduces new hazards you should consider them and, if significant, do whatever you need to do to keep the risks under control. In any case you should keep your assessment under review to make sure that the precautions are still working effectively. You may want to re-examine the fire prevention and protection measures at the same time as your health and safety assessment.

If a fire or 'near miss' occurs, this could indicate that your existing assessment may be inadequate and you should carry out a re-assessment. It is good practice to identify the cause of any incident and then review and, if necessary, revise your fire risk assessment in the light of this.

Records of testing, maintenance and training etc. are useful aids in a review process. See Appendix A.1 for an example.

Alterations notices

If you have been served with an 'alterations notice' check it to see whether you need to notify the enforcing authority about any changes you propose to make as a result of your review. If these changes include building work, you should also consult a building control body.

END OF PART 1

You should now have completed the five-step fire risk assessment process, using the additional information in Part 2 where necessary. In any review you may need to revisit Steps 1 to 4.

Part 2 Further guidance on fire risk assessment and fire precautions

Managing fire safety

Good management of fire safety in your premises is essential to ensure that any fire safety matters that arise are always effectively addressed. In simple premises this can be achieved by the manager or owner maintaining and planning fire safety in conjunction with general health and safety.

In larger premises or organisations it is good practice for a senior manager to have overall responsibility for fire safety. It may be appropriate for this responsibility to be placed with the manager designated with overall responsibility for health and safety.

An organisation's fire safety policy should be flexible enough to allow modification. This is particularly important when local managers have to function daily with other businesses in the same building. It should be recognised that fire safety operates at all levels within an organisation and therefore local managers should be able to develop, where necessary, a local action plan for their premises.

The organisation's policy should be set out in writing and may cover such things as:

- who will hold the responsibility for fire safety at board level;

- who will be the responsible person for each of their premises (this will be the person who has overall control; usually the manager, but may be part-time or shift managers);

- the arrangement whereby managers will, where necessary, nominate in writing specific people to carry out tasks if there is a fire; and

- the arrangement whereby regional or area managers should monitor and check that individual managers are meeting the requirements of the fire safety law.

You should have a plan of action to bring together all the features you have evaluated and noted from your fire risk assessment so that you can logically plan what needs to be done. It should not be confused with the emergency plan, which is a statement of what you will do if there is a fire.

The plan of action should include what you intend to do to reduce the hazards and risks you have identified and to implement the necessary protection measures.

You will need to prioritise these actions to ensure that any findings which identify people in immediate danger are dealt with straight away, e.g. unlocking fire exits. In other cases where people are not in immediate danger but action is still necessary, it may be acceptable to plan this over a period of time.

You (or a designated manager) should be in charge of and be present on the premises during the whole time that the premises are occupied by residents, and this designated person should be kept free from work which would prevent them from being available immediately in the event of an alarm of fire.

It is important that the manager ensures that where, for operational reasons, staff are transferred to another part of the premises, they are made aware of the means of escape and fire procedures of the new work area (including any additional responsibilities) if this differs from their permanent duty station.

The fire and rescue service should be called to every outbreak of fire.

The guidance in Part 2 provides additional information to:

- ensure good fire safety management by helping you establish your fire prevention measures, fire precautions and fire safety procedures (systems equipment and plans); and

- assist you to carry out your fire safety risk assessment, and identify any issues that need attention.

Section 1 Further guidance on fire risks and preventative measures

This section provides further information on evaluating the risk of a fire and its prevention in your premises. You should spend time developing long-term workable and effective strategies to reduce hazards and the risk of a fire starting. At its simplest this means separating flammable materials from ignition sources.

You should consider:

- housekeeping;
- laundries;
- kitchen areas;
- storage;
- dangerous substances: storage and use;
- equipment and machinery;
- electrical safety;
- smoking;
- managing building work and alterations;
- existing layout and construction;
- particular hazards in corridors and stairways used as escape routes;
- insulated core panels;
- restricting the spread of fire and smoke;
- arson; and
- help for people with special needs.

1.1 Housekeeping

For all care homes, good housekeeping will lower the chances of a fire starting, so the accumulation of combustible materials in premises should be monitored carefully. Good housekeeping is essential to reduce the chances of escape routes and fire doors being blocked or obstructed.

Keep waste material in suitable containers before it is removed from the premises. If bins, particularly wheeled bins, are used outside, secure them in a compound to prevent them being moved to a position next to the building (see Figure 13) and set on fire. Never place skips against a building – they should normally be a minimum of 6m away from any part of the premises.

If you generate a considerable quantity of combustible waste material then you may need to develop a formal plan to manage this effectively.

Other housekeeping issues include:

- the appropriate storage of aerosol sprays and medical gases;
- avoiding the use of flammable materials and liquids; and
- orderly stacking of linen, paper, medical consumables, packaging, furniture and furnishings.

In premises that provide care, the predictable nature of the day-to-day activities should allow systems to be developed for dealing with waste, laundry and other combustible materials.

Such waste should not be allowed to accumulate inside the building, particularly in the escape routes, whilst awaiting collection. Consideration of this issue should form part of the building fire risk assessment.

In higher risk areas you need to make sure arrangements are in place for safe close down, e.g. checking all appliances are turned off and combustible waste has been removed.

Figure 13: Bins under stairway (courtesy of Cheshire fire service)

1.2 Laundries

Laundries, in both large and small care premises, remain a high-risk area. They are often located in the basement, which means that any fire can affect the escape routes above.

Washing and drying machines should not be loaded in excess of the manufacturer's recommendations, exhaust filters should be cleaned, and maintenance carried out regularly. Items such as cleaning cloths and mop heads placed in the dryers can spontaneously combust if there is any chemical residue left on them.

Ironing equipment should be correctly used and maintained. The laundry area should not be used for storing miscellaneous combustible material.

The use of laundry chutes in larger premises provides a ready path for smoke from any fire to travel throughout the residential areas of the premises. Any smoke and fire dampers within the laundry chute should be automatically operated following activation of fire detection devices within the chute.

1.3 Kitchens

In larger premises with extensive catering facilities the cooking range should have some form of automatic fire suppression system. In smaller premises a suitable fire extinguisher and fire blanket should always be provided. All deep fat cooking equipment should have a thermostatic temperature control and should never be left unattended. Open cooking, such as frying, should not be left unattended. In some premises, such as a small children's home, the use of 'open top chip pans' should be discouraged or prohibited. There are approximately 8,000 chip pan fires every year.*

Extractor ducting, grease traps and filters should be regularly cleaned and maintained. Isolation switches for gas and electricity supplies, and isolation switches for any extractor fans should be located near to an exit.

1.4 Storage

Many of the materials found in your premises will be combustible. If your premises have inadequate or poorly managed storage areas then the risk of fire is likely to be increased. The more combustible materials you store the greater the source of fuel to a fire. Poorly configured storage could prevent equipment such as sprinklers working effectively.

In premises used for residential care there can often be quite bulky (combustible) equipment, such as wheelchairs and bath chairs, which need to be stored away when not in use.

Combustible materials are not just those generally regarded as highly combustible, such as polystyrene, but all materials that will readily catch fire. However, by carefully considering the type of material, the quantities kept and the storage arrangements, the risks can be significantly reduced.

Figure 14: Poor storage

In your office (if you have one), the retention of large quantities of paper records, especially if not filed away in proprietary cabinets, can increase the fire hazard. Such readily available flammable material makes the potential effect of arson more serious.

*Fire Statistics UK 2004: ODPM 2006

> **Case study**
>
> An old people's home used the space beneath an accommodation stairway to store a large number of bath chairs. This was a substantial fire load.
>
> The area was formed into a protected area by constructing a fire-resisting wall and by underdrawing the stairs. Fire doors were fitted to allow the chairs to be moved in and out easily.

Many premises will take great care to present an efficient and attractive image in the areas used by residents and visitors, while storage areas are neglected and allowed to become dumping areas for unneeded material. Do not pile combustible material against electrical equipment or heaters, even if turned off for the summer, and do not allow smoking in areas where combustible materials are stored.

To reduce the risk, store excess combustible materials and stock such as baggage, linen and furniture in a dedicated storage area, storeroom or cupboard that is fire-resisting. Do not store excess stock in escape routes or areas where residents or visitors would normally have access. Bathrooms should not be used for storage unless they are separated from the rest of the premises by fire-resisting doors and walls, and are fitted with automatic fire detection.

Your fire risk assessment should also consider any additional risk generated by seasonal products such as fireworks and Christmas decorations.

Consider the following to reduce these risks:

- use fire-retardant materials wherever possible (suppliers should be able to provide evidence of this);
- ensure electrical equipment in your store rooms does not become a potential source of ignition;
- provide adequate space for equipment, furniture and wheelchairs;
- allow storage in designated areas only; and
- provide separate storage areas for medical gases, aerosols and flammable gases (see Section 1.5).

Voids

Voids (including roof voids) should not be used for the storage of combustible material. Such voids should be sealed off or kept entirely open to allow for easy access for inspection and the removal of combustible materials.

1.5 Dangerous substances: storage and use

Specific precautions are required when handling and storing dangerous substances to minimise the possibility of an incident. Your supplier should be able to provide detailed advice on safe storage and handling. However, the following principles will help you reduce the risk from fire:

- substitute highly flammable substances and materials with less flammable ones;
- reduce the quantity of dangerous substances to the smallest reasonable amount necessary for running the business or organisation;
- correctly store dangerous substances, e.g. in a fire-resisting enclosure. All flammable liquids and gases should ideally be locked away, especially when the premises are unoccupied, to reduce the chance of them being used in an arson attack; and
- ensure that you and your staff are aware of the fire risk the dangerous substances present and the precautions necessary to avoid danger.

You are not likely to have very large quantities of such materials, but sometimes this can happen, for example after you have re-stocked with aerosol products (e.g. air-fresheners), or during room redecoration. You may keep a stock of petrol for a lawn mower, or oxygen, for medical purposes.

Additional general fire precautions may be needed to take account of the additional risks that may be posed by the storage and use of these substances.

Certain substances and materials are by their nature, highly flammable, oxidising or potentially explosive. These substances are controlled by other legislation in addition to fire safety law, in particular the Dangerous Substances and Explosive Atmospheres Regulations 2002[7] (also see HSE's *Approved Code of Practice and guidance*[8]).

Flammable liquids

Highly flammable liquids present a particularly high fire risk. For example, a leak from a container of flammable solvents, such as acetone, may produce large quantities of heavier-than-air flammable vapours. These can travel large distances, increasing the likelihood of their reaching a source of ignition well away from the original leak, such as a basement containing heating plant and/or electrical equipment on automatic timers.

Flammable liquids stored in plastic containers can be a particular problem if involved in fire because they readily melt spilling their contents and fuelling rapid fire growth.

The risk is reduced by ensuring the storage and use of highly flammable liquids is carefully managed, that materials contaminated with solvent are properly disposed of and when not in use, they are safely stored. Up to 50 litres may be stored in a fire-resisting cabinet or bin that will contain any leaks.

Figure 15: A fire-resisting pedal bin for rags

Figure 16: A 50 litre flammable storage bin

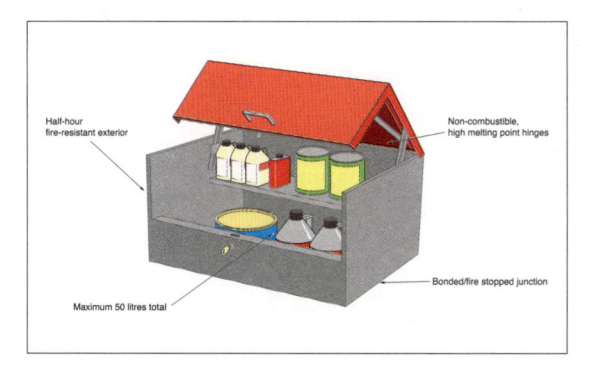

There should be no potential ignition sources in areas where inflammable liquids are used or stored and flammable concentrations of vapours may be present. Any electrical equipment used in these areas, including fire alarm and emergency lighting systems, needs to be suitable for use in flammable atmospheres. In such situations you should seek advice from a competent person.

LPG storage and use

Liquefied petroleum (bottled) gas (LPG) appliances and cylinders and, in particular, portable LPG equipment, should only be used in care homes in exceptional circumstances. They should not be used or allowed in situations where they cannot be adequately and safely controlled. Nor should they be used where the use of mains gas is also prohibited.

Only butane cylinders should be used inside residential care premises, except when propane is required for building or maintenance work. Cylinders should not be used, or stored, in basements or cellars which have no natural floor level ventilation. Where it is necessary to keep spare butane cylinders indoors, the amount should be limited to a maximum of 15kg in not more than two cylinders. These should be kept in a safe, secure and well ventilated place where:

- they cannot be interfered with;
- can be kept upright (with valve protection fitted);
- are away from sources of ignition and/or readily ignitable materials; and
- are away from any corrosive, toxic or oxidant materials.

Bulk storage tanks for LPG should be designed, installed and located in accordance with industry guidance.

Your supplier should be able to provide further guidance on the safe storage and use of LPG. Detailed advice is also given by the Liquefied Petroleum Gas Associations Codes of Practice.[9,49,50]

Advice on the use of LPG for heating is given below (Section 1.6).

Aerosols

Some aerosol cans contain flammable products stored at pressure. They can present a high hazard. When ignited, they can explode and produce fireballs. Their presence can make it unsafe for firefighters to enter a building and they have the potential for starting multiple fires.

Cleaning and other staff should be made aware of the potential dangers of aerosols and the need for their safe disposal.

Damaged and leaking aerosols should be removed immediately to a safe secure well ventilated place.

> **Case study**
>
> A can of aerosol polish was left on the window sill of a kitchen in a care home. The heat from the sun caused the aerosol to burst releasing a flammable gas cloud into the kitchen which was ignited by a lit gas ring causing an explosion which blew out the window and two internal doors. It is vital to ensure that dangerous substances are stored safely, preferably in secure areas and always in accordance with the manufacturers instructions.

Medical gases

Where medical oxygen is being used, you should consider the following:

- the room where the oxygen is used should be a designated a 'No smoking area';

- there should be clear instructions and warnings for occupants – in particular 'no smoking';

- the room should be adequately ventilated to ensure any leaks of oxygen cannot accumulate;

- all staff involved in operating the oxygen equipment should be properly trained, including the provision of written procedures; and

- the equipment should be designed and constructed for use with oxygen [further advice on this may be obtained from the Medical and Healthcare Products Regulatory Agency (MHRA)* or from your supplier.

*mhra.gov.uk

The number of medical oxygen cylinders stored should be kept to the minimum necessary to ensure sufficient supply is available for patients needs. Where possible, especially when significant numbers of cylinders need to be stored, the cylinders should be preferably stored outside, in a safe and secure location where they cannot be interfered with.

Where spare cylinders need to be stored indoors, the number should kept to the practical minimum and located in well ventilated areas, but not in passageways, stairwells or adjacent to emergency exits. Ideally, they should be located in designated marked cupboards or rooms provided with permanent ventilation to the outside.

Cylinders should be:

- kept away from extremes of heat, fires and naked lights (smoking should be prohibited in the vicinity);

- not be stored areas where there is a possibility of their being contaminated with oils or greases; e.g. kitchen and garage areas;

- secured to prevent their falling over; and

- separated, full from empty cylinders, to avoid confusion.

The fire and rescue service should be informed if oxygen is used and stored on the premises.

Further advice is given by the European Industrial Gases Association.[52]

Hazardous materials

Where it is necessary to use or store hazardous materials, such as fuels (whether in containers or within fuel tanks and machinery), fertilisers, weed killers, paints and solvents, they should be stored in a secure and safe location, for example a properly ventilated fire-resisting bin or storeroom. Care should also be taken to ensure that incompatible materials are not stored together. If in any doubt you should seek the advice of your supplier.

1.6 Equipment and machinery

In residential care homes, equipment and machinery is most commonly found in those with day care facilities, laundries or workshops. Lack of preventive maintenance increases the risk of fire starting in machinery. Common causes of fire in equipment are:

- allowing ventilation points to become clogged or blocked, causing overheating;
- inadequate cleaning of equipment, such as tumble driers;
- allowing extraction equipment in kitchens to build up excessive grease deposits;
- loose drive belts or lack of lubrication leading to increased friction;
- disabling or interfering with automatic or manual safety features and cut-outs;
- in workshops, build-up of saw dust; and
- leaking valves, glands or joints allowing oils and other flammable liquids to contaminate adjacent floors or goods.

A competent person should adequately and regularly maintain machinery, equipment and plant, including cooking and heating equipment, packaging apparatus and office equipment such as photocopiers. Appropriate signs and instructions on safe use may be necessary.

Heating

Individual heating appliances require particular care if they are to be used safely, particularly those which are kept for emergency use during a power cut or as supplementary heating during severe weather. The greatest risks arise from lack of maintenance and staff unfamiliarity with them. All heaters should preferably be secured in position when in use and fitted with a fire guard if appropriate.

As a general rule, convector or fan heaters should be preferred to radiant heaters because they present a lower risk of fire and injury. The following rules should be observed:

- All heaters should be kept well clear of combustible materials and in a position where they do not cause an obstruction.
- Heaters which burn a fuel should be sited away from draughts.
- Portable fuel burning heaters, including bottled gas (LPG), should only be used in areas used by residents or visitors in exceptional circumstances and if shown to be acceptable in your risk assessment. (See Section 1.5 above.)

All gas heating appliances should be used only in accordance with manufacturer's instructions and should be serviced annually by a competent person.

In general, residents, visitors or staff should be discouraged from bringing their own portable heaters into the premises.

The use of boiler rooms to store combustible materials should be avoided.

1.7 Electrical safety

Electrical equipment can be a significant cause of accidental fires in buildings. The main causes are:

- overheating cables and equipment, e.g. due to overloading circuits, bunched or coiled cables or impaired cooling fans;
- incorrect installation or use of equipment;
- damaged or inadequate insulation on cables or wiring;
- combustible materials being placed too close to electrical equipment which may give off heat even when operating normally or may become hot due to a fault;
- arcing or sparking by electrical equipment;
- overloading of sockets;
- inadequate servicing and maintenance of equipment, e.g. electric blankets;
- overheating cables due to retention in a coil; and
- lack of maintenance or testing.

All electrical equipment should be installed and maintained in a safe manner by a competent person. If portable electrical equipment is used, including items brought into a workplace by staff, then your fire risk assessment should ensure that it is visually inspected and undergoes portable appliance testing ('PAT') at intervals suitable for the type of equipment and its frequency of use (refer to HSE guidance[10]). If you have any doubt about the safety of your electrical installation then you should consult a competent electrician.

Issues to consider include:

- overloading of equipment;
- correct fuse ratings;
- PAT testing and testing of the fixed installation;

- protection against overloading of installation;
- protection against short circuit;
- insulation, earthing and electrical isolation requirements;
- frequency of electrical inspection and test;
- temperature rating and mechanical strength of flexible cables;
- portable electrical equipment, including lanterns;
- physical environment in which the equipment is used (e.g. wet or dusty atmospheres); and
- suitable use and maintenance of personal protective equipment.

All electrical installations should be regularly inspected by a competent electrical engineer appointed by you, or on your behalf, in accordance with the Electricity at Work Regulations.[53] The use of low voltage equipment should conform to the requirements of the Electrical Equipment (Safety) Regulations 1994,[54] including the requirement to be CE marked.

1.8 Smoking

Carelessly discarded cigarettes and other smoking materials are a major cause of fire. A cigarette can smoulder for several hours, especially when surrounded by combustible material. Many fires are started several hours after the smoking materials have been emptied into waste bags and left for future disposal.

The control of smoking in care homes is very important particulary where certain types of resident are concerned. For example, confusion in older people may lead to the careless disposal of cigarettes. Smoking policies must be practical but must also recognise the human rights of residents to enjoy their preferred lifestyle. Your risk assessment must take account residents who wish to smoke in their own rooms or smoking residents who are confined to bed and can not access designated smoking areas.

Consider operating a safe smoking policy in designated smoking areas and prohibit smoking elsewhere. Display suitable signs throughout the premises informing people of the smoking policy and the locations where smoking is permitted. Prohibit smoking in fire hazard rooms (e.g. laundries and storerooms) and protected routes (e.g. lobbies, stairways) and ensure that your policy is enforced.

In those areas where smoking is permitted, provide deep and substantial ashtrays to help prevent unsuitable containers being used. Empty all ashtrays daily into a metal waste bin and take it outside. It is dangerous to empty ashtrays into plastic waste sacks which are then left inside for disposal later. You should carry out regular inspections in all areas, particularly smoking permitted areas once these areas have been vacated at night.

1.9 Managing building work and alterations

Fires are more frequent when buildings are undergoing refurbishment or alteration.

You should ensure that, before any building work starts, you have reviewed the fire risk assessment and considered what additional dangers are likely to be introduced. You will need to evaluate the additional risks to people, particularly since your care home may continue to be occupied. Lack of pre-planning can lead to haphazard co-ordination of fire safety measures.

You should liaise and exchange information with contractors who will also have a duty under the Construction (Health, Safety and Welfare) Regulations 1996[11,12] to carry out a risk assessment and inform you of their significant findings and the preventive measures they may employ. This may be supported by the contractors' agreed work method statement. The designer should also have considered fire safety as part of the Construction (Design and Management) Regulations 1994 (the CDM Regulations).[55]

You should continuously monitor the impact of the building work on the general fire safety precautions, such as the increased risk from quantities of combustible materials and accumulated waste. You should only allow the minimum materials necessary for the work in hand within or adjacent to your building.

Additional risks can include:

- hot work such as flame cutting, soldering, welding or paint stripping;
- temporary electrical equipment;

- blocking of escape routes including external escape routes;
- introduction of combustibles into an escape route;
- loss of normal storage facilities;
- fire safety equipment, such as automatic fire detection systems becoming affected;
- fire-resisting partitions being breached or fire doors being wedged open (see Appendix B1 for information on fire-resisting separation); and
- additional personnel who may be unfamiliar with the premises.

Activities involving hot work such as welding, flame cutting, use of blow lamps or portable grinding equipment can pose a serious fire hazard and need to be strictly controlled when carried out in areas near flammable materials. This can be done by having a written permit to work for the people involved (whether they are your employees or those of the contractor).

A permit to work is appropriate in situations of high hazard/risk and, for example, where there is a need to:

- ensure that there is a formal check confirming that a safe system of work is being followed;
- co-ordinate with other people or activities;
- provide time-limits when it is safe to carry out the work; and
- ensure that specialised personal protective equipment (such as breathing apparatus) has been provided and/or that adequate methods of communication are available.

Further guidance on fire safety during construction work is available from the HSE[56,57] and the Fire Protection Association.[58]

1.10 Existing layout and construction

In many premises, the design is for open-plan areas, such as recreation areas, allowing residents and staff to move freely throughout the area.

Traditionally, occupants are advised to shut doors when escaping from a fire in order to contain the fire and smoke and stop it spreading. In open-plan areas the fire, and especially the smoke, may spread unimpeded, and it may spread faster than expected.

To assess the risk in your premises you need to evaluate the construction and layout of your premises. This does not mean a structural survey, unless you suspect that the structure is damaged or any structural fire protection is missing or damaged, but rather an informed look around to see if there are any easy paths through which smoke and fire may spread and what you can do to stop that. In general, older buildings will have more void areas, possibly hidden from view, which will allow smoke and fire to spread away from its source. Whatever your type of building, you may need to consider typical situation that may assist the spread of fire and smoke such as:

- vertical shafts, e.g. lifts, open stairways, dumb waiters, and linen and waste chutes;
- false ceilings, especially if they are not fire-stopped above walls;
- voids behind wall panelling;
- unsealed holes in walls and ceilings where pipe work, cables or other services have been installed;
- doors, particularly to stairways, which are ill-fitting or routinely left open; and
- storage under stairs

1.11 Particular hazards in corridors and stairways used as escape routes

Items that are a source of fuel, pose an ignition risk, or are combustible and likely to increase the fire loading or spread of fire, should not be located on any corridor or stairway or circulation space that will be used as an escape route. Such items include:

- portable heaters, e.g. bottled gas (LPG) or electric radiant heaters and electric convectors or boilers;
- gas cylinders for supplying heaters;
- cooking appliances; and
- unenclosed gas pipes, meters, and other fittings.

However, depending on the findings of your risk assessment and, where more than one escape route is available, items such as those below may be acceptable if the minimum exit widths are maintained and the item presents a relatively low fire risk:

- non-combustible lockers;
- vending machines; and
- small items of electrical equipment (e.g. photocopiers).
- small areas of clothing and/or small quantities of upholstered furniture which meet BS 7176[48] or the Furniture and Furnishing (Fire) (Safety) Regulations 1988 (as amended 1989 and 1993).[94]

1.12 Insulated core panels

Some buildings used as care homes, or as part of a care home, have insulated core panels as exterior cladding or, more likely, for internal structures and partitions. Insulated core panels are easily constructed, enabling alterations and additional internal partitions to be erected with minimum disruption to daily life.

They normally consist of a central insulated core, sandwiched between an inner and outer metal skin coated with a PVC covering. The central core can be made of various insulating materials, ranging from virtually non-combustible through to highly combustible. Fire can grow unseen inside the panel and only become obvious when it is well developed.

Figure 17: Insulated core panels – internal panel

It is difficult to identifty the type of core the panels have, therefore best practise can help you reduce any additional risk:

- Do not store highly combustible materials or rubbish, or install heating appliances against the panels.
- Control ignition sources that are adjacent to, or penetrating the panels.
- Have damaged panels or sealed joints repaired immediately and make sure that jointing compounds or gaskets used around the edges of the panels are in good order.
- Check where openings have been made for doors, windows, cables and ducts that these have been effectively sealed and the inner core has not been exposed.
- Ensure there has been no mechanical damage, e.g. caused by mobile equipment, such as bath chairs.
- Ensure that the panels are correctly secured to the structure or are designed to be independently structurally secure. One solution is to ensure the retaining brackets bolt the panel to a secure frame through the outer and inner skins.
- Ensure that any loads, such as storage and equipment, are supported by such panels only if they have been designed and installed to perform this function.

The use of combustible panels in your care home should be carefully considered. Your fire risk assessment may need to be revised to ensure that any increased risk resulting from this type of construction is considered.

Insulated core panels should be installed by a competent person in accordance with industry guidance.

Guidance on the design, construction, specification and fire management of insulated core panels has been established by the International Association of Cold Storage Contractors.[59]

1.13 Restricting the spread of fire and smoke

To reduce the risk to people if there is a fire, you need to consider how to control or restrict the spread of fire and smoke. The majority of people who die in fires are overcome by the smoke and gases. It is important therefore to ensure that, in the event of fire, the rate of fire growth is restricted in its early stages. Most measures which restrict the rate of fire growth in its early stages will also serve to restrict the fire spread in its later stages.

Furniture (including beds), soft furnishings and textiles

The type and quantity of any combustible contents stored or used in your premises can significantly affect the way a fire can grow and spread therefore, using only products that are resistant to ignition and flame spread can significantly help to reduce the fire risk to people.

Upholstered furniture, soft furnishings and textiles which are easily ignited or have rapid spread of flame characteristics should, where practicable, be avoided in any type of care home and all such products should meet a recognised fire performance standard.

Information on fire performance and testing of furniture, furnishings and textiles can be found in the following standards:

- Ignitability of upholstered furniture (including composites of cover material and infill) and loose covers: BS 5852.[61]
- Resistance to ignition of upholstered furniture for non-domestic use: BS 7176.[48]
- Resistance to ignition of mattresses, divans and bed bases: BS 7177.[86]
- Ignitability of mattresses with waterproof covers: BS 6807,[69] Section 2.
- Flammability of fabrics for curtains and drapes (including nets and linings, blackout curtains and roller blinds): BS 5867-2,[62] Type, Type B.
- Burning behaviour (ignitability and flame spread) of curtains and drapes: BS EN 1101[63] (ignitability) and BS EN 1102[64] (flame spread).

Note: For all the above standards equivalent European classifications and standards, where available, may also be used.

Additional guidance on the fire performance of textiles and furniture can be found in the Firecode document HTM 87.[60]

In some existing premises you may not be able to achieve appropriate fire performance standards immediately, but they should be applied to all new or replacement items. Where your fire risk assessment identifies significant quantities of furniture and furnishings that do not meet an appropriate standard of fire performance you will need to consider what alternative fire precautions may need to be put in place to compensate.

Textile fabrics that are capable of meeting an appropriate standard of fire performance include durably flame retardant treated cotton and 100% flame retardant polyester. 100% modacrylic is also suitable if available.

It is important to note that retaining the flame retardant effectiveness of all chemically-treated flame retardant fabrics, such as flame retardant treated cotton, is dependent on the correct laundry procedures being followed. Careful note should be taken of all wash and care instructions provided. Advice should be sought from the supplier if no such instructions are provided.

If in any doubt about the fire performance of any flame retardant treatment, product or material you should seek confirmation from the supplier that the items have been tested for fire performance flammability by a UKAS* accredited organisation.

Residents should only be allowed to provide items of their own furniture or textiles for their own bedroom if the item meets a fire performance standard similar to those described above, particularly if they share the bedroom with other residents.

You should take care when selecting decorative finishes or multi-layer decorative systems, e.g. textile based linings and oil based paints, particularly those used for wall and ceiling linings, and ensure that they are suitably fire retardant. Further information about the surface finishes of wall and ceiling linings is available in Appendix B1

*United Kingdom Accreditation Service. www.ukas.com

Textile floor coverings are not normally a serious fire hazard, but all soft floor coverings should be resistant to ignition and flame spread. Textile floor coverings bonded to the floor present a lower fire risk than those loosely laid. BS 5287[51] gives a specification for the assessment and labelling of textile floor coverings.

Polypropylene chairs should have flame retardant polypropylene shells.

Soft play environments, although not furniture as such, can contain a large volume of covered plastic foam in various shapes. The covered foam should be at least as resistant to ignition as upholstered furniture (see reference to upholstered furniture above). Where such environments are extensive, as in play rooms, extra care must be taken and additional precautions may be necessary.

Bedding and sleepwear
Bedding and sleepwear should also meet recognised standards of fire performance. Information on fire performance specifications for bedding and sleepwear can be found in the following standards:

- Flammability of blankets for use in the public sector: BS 5866-4,[65] or for ignitability of a representative fabric sample: BS 7175[67]

- Flammability of counterpanes for use in the public sector: BS 5815-3.[68]

Note: Counterpanes covering the bed at all times will help to protect the other textile bedding items from involvement in a fire.

- Ignitability of continental quilts and duvets: BS 7175.[67]

- Flammability of quilt covers: BS 5815-3.[68]

- Ignitability of bed covers and pillows: BS 7175[67] Section 4.

- Flammability of sleepwear (including dressing gowns and bed jackets): BS 5722.[70]

Note: This standard gives three different levels of flammability performance. Level 3 is suitable for normal situations, but higher performance levels can be specified for high risk situations (level 1 being the highest).

Note: For all the above standards equivalent European classifications and standards, where available, may also be used.

Bedding and sleepwear provided by the residents or their families should, as far as is practicable, meet equivalent standards of fire performance.

Many products such as mattress overlays, fleeces and underpads are used in the care of residents with, or with a pre-disposition to, pressure sores. They are usually placed within the bed assembly, or on easy chairs or wheelchairs. Where possible these products should be resistant to ignition. BS 7175,[67] Section 3 gives further guidance. However, nursing or medical advice should be sought if there is likely to be a conflict between a resident's needs and fire safety.

As with soft furnishings, it is important to note that retaining the flame retardant effectiveness of all chemically-treated flame retardant fabrics, such as flame retardant treated cotton, is dependent on the correct laundry procedures being followed. Careful note should be taken of all wash and care instructions provided. Advice should be sought from the supplier if no such instructions are provided.

Electric blankets should be PAT tested and only used in accordance with manufacturer's instructions.

Toys
Soft toys should comply with the flammability retardancy requirements of the Toys (Safety) Regulations 1995.[71] The care and cleaning of any soft toys should be in accordance with the manufacturer's instructions in order to maintain these standards. flame retardancy standards. You should carefully check any soft toys donated to the home to ensure that they also meet the requirements of the Regulations.

Fire-resisting structures
Many buildings are divided into different areas by fire-doors and fire resisting walls and floors. These are partly designed to keep a fire within one area, giving people more time to escape. You will need to identify which doors, walls and floors in your building are fire resisting. There may be information available from when the building was built, if alterations have been made, or from a previously held fire certificate.

Normally if there are fire-doors in a wall, then the wall itself will also need to be fire resisting. (See Appendix B1 for more technical information about fire-resisting walls and doors). If you decide that a wall or floor is required to be fire resisting then you should not make any holes in it, e.g. for extra doors or pipe ducts, without consulting a competent person.

Smoke control arrangements

In larger buildings there may be some form of smoke control provided for the safety of the occupants and to assist fire-fighting (e.g. Smoke and Heat Exhaust Ventilation Systems (SHEVS), openable vents (OVs), or automatic opening vents (AOVs)). These systems are designed to restrict the spread of fire and smoke usually by venting the heat and smoke through the roof or via other routes to outside. Low level inlet air is essential for the operation of SHEVS and all openings for this purpose should not be obstructed.

It is important that any smoke can flow easily into the reservoirs and that nothing which could cause an obstruction, e.g. stored mattresses, is fixed near the vents.

If your building has smoke vents fitted, or any other form of smoke control, then you may need to seek advice from someone who is competent in such systems. Further information on smoke control can be found from CIBSE Guide E[72] or from the BRE.[73]

Ventilation systems

Where ventilation systems might assist the spread of flames, smoke and hot gases from a fire it will be necessary to take steps to safeguard the means of escape against this hazard. Ventilation ducts should be fitted with fire dampers (which close in the presence of fire) where they cross compartment boundaries (walls or floors) but may also need smoke dampers which operate on actuation of the fire warning system.

Sprinklers

In some premises there may be a sprinkler system installed. Sprinklers are designed to restrict the spread of fire by suppressing the fire in its early stages. Further guidance on sprinklers can be found in Part 2, Section 3.2.

1.14 Arson

Recent studies indicate that, over 2,100 serious deliberately set fires, resulting in two deaths and 55 injuries, occur every week.* All premises can be targeted either deliberately or just because they offer easy access.

Be aware of other small, deliberately set fires in the locality, which can indicate an increased risk to your premises. Be suspicious of and record any small 'accidental' fires on the premises and investigate them fully.

Fires started deliberately can be particularly dangerous because they generally develop much faster and may be intentionally started in escape routes. Of all the risk-reduction measures, the most benefit may come from efforts to reduce the threat from arson.

Measures to reduce arson may include the following:

- ensure the outside of the premises is well lit and, if practical, secure the perimeter of the premises;
- thoroughly secure all entry points to the premises, including windows and the roof, but make sure that this does not compromise people's ability to use the escape routes;
- make sure you regularly remove all combustible rubbish;
- do not place rubbish skips adjacent to the building and secure waste bins in a compound separated from the building;
- do not place vehicles, caravans or other portable structures adjacent to the building;
- encourage staff to challenge people acting suspiciously;
- remove automatic entry rights from staff who have been dismissed;
- ensure that your security alarm/fire-detection system is monitored and acted on;
- secure flammable liquids so that intruders cannot use them;
- secure all storage areas and unused areas of the building that do not form part of an escape route against unauthorised access, ensure access to keys to those areas is restricted;
- fit secure metal letterboxes on the inside of letter flaps to contain any burning materials that may be pushed through;
- deter unauthorised entry to the site;
- prevent unauthorised entry into the building;
- reduce the opportunity for people with mental health/learning needs to start a fire;
- maintain security of the main access door in the event of a fire.

Further guidance on reducing the risk of arson has been published by the Arson Prevention Bureau.**

*Fire Statistics UK, 2004: DCLG 2006.
**Visit www.arsonpreventionbureau.org.uk for more information.

1.15 Help for people with special needs

By their nature, premises that provide care will often have residents who, in addition to being elderly, or very young, or in need of specialist care, will also have some other disability. The Disability Rights Commission estimates that 11 million people in this country have some form of disability. This may impact on their ability to leave a building speedily in the event of fire or make them entirely dependant on others to escape.

Some common forms of disability that you may need to take account of in your risk assessment include:

- mobility impairment, which can limit speed of evacuation;
- hearing impairment, which can limit the response to an alarm;
- visual impairment, which can limit the ability to escape; and
- learning difficulties, which can affect the response to an alarm.

As a person responsible for providing care services you are likely to have considerable experience in assisting disabled people to move about. You should therefore carefully assess the practicalities of undertaking an emergency evacuation in the event of fire. If you have disabled employees you may also need to discuss their individual needs.

Under the Disability Discrimination Act, if disabled people could realistically expect to use the service you provide then you must anticipate any reasonable adjustments that would make it easier for that right to be exercised. Accordingly, if disabled people are going to be in your premises then you must also provide a safe means for them to leave if a fire occurs.

The Disability Discrimination Act includes the concept of 'reasonable adjustments' and this can be carried over into fire safety law. It can mean different things in different circumstances. For example, in a small children's home it may be considered reasonable to provide contrasting colours on a handrail to help those with vision impairment to follow an escape route more easily. However, it might be unreasonable to expect that same home to install an expensive voice alarm system. Appropriate 'reasonable adjustments' for a large business or organisation may be much more significant.

In premises with many severely disabled residents, you may also wish to contact a professional access consultant or take advice from disability organisations.

Whilst many people with special needs wish to and are able to facilitate their own escape, there may be a significant number of people in premises that provide care who are severely disabled and only able to move or react adequately with assistance from carers or staff. These will include residents who are confined to bed and receiving medical interventions by way of attached medical devices.

You may conclude that the current levels of assistance available in your premises, and the layout and construction means that the evacuation of some people (most likely residents) cannot be guaranteed within an acceptable time. You will then need to consider some additional method of ensuring their safety (e.g. an automatic fire suppression system). In such cases you should seek specialist advice from a competent person.

Where people with special needs (residents, employees and visitors) are accommodated, work in, or use the premises, their needs should, so far as is practicable, be discussed with them. These will often be modest and may require only changes or modifications to existing procedures. However, in some cases, more individual arrangements involving the development of 'personal emergency evacuation plans' (PEEPs) may need to be considered. Any PEEP developed for residents may need to be incorporated into the inviduals care plan.

Guidance on removing barriers to the everyday needs of disabled people is contained in BS 8300[14]. Much of this advice will also assist disabled people during an evacuation.

You should ensure that your emergency plan has a record of where disabled people are located in the building and includes a plan of action to assist them in the event of a fire.

Further advice can be obtained from the Disability Rights Commission at www.drc-gb.org.

Section 2 Further guidance on fire detection and warning systems

Automatic fire detection (AFD) and warning systems provide the means to notify you of the presence of fire at the earliest possible opportunity. It offers you the possibility of carrying out firefighting (because the fire is still small) and the maximum period of time to implement your emergency plan and to evacuate your residents and staff. The AFD system can be linked to other active fire safety systems in your building (e.g. door closing devices and smoke control vents) so that they operate automatically.

Your electrical fire detection and warning system should typically include the following:

- automatic fire detectors, e.g. smoke, heat;
- manual call points (break-glass call points) next to exits with at least one call point on each floor;
- electronic sirens or bells; and
- a control and indicator panel.

If, for any reason, your system fails you must still ensure that people in your premises can be warned and escape safely. A temporary arrangement, such as gongs, whistles or air horns, combined with suitable training, may be acceptable for a short period.

The fire warning sound levels should be loud enough to alert everyone, taking into account background noise. In areas with high background noise, or where people may be wearing hearing protectors, the audible warning should be supplemented, e.g. with visual alarms. In sleeping areas, to ensure that people are woken, a sufficient sound level should be achieved at the head of the bed (i.e. 75dBA). This will usually mean fitting a sounder in each bedroom. Where this is not desirable, e.g. because this might cause alarm to people who are dependant on staff to help them escape, then a sound level of at least 45dBA should be available throughout the premises to alert staff and others who are awake. Where staff sleep on the premises, a sound level of 75dBA should be provided at the bed head in staff bedrooms.

People with hearing difficulties

Where people have hearing difficulties, particularly those who are profoundly deaf, then simply hearing the fire warning is likely to be the major difficulty. If these persons are never alone while on the premises then this may not be a serious problem, as it would be reasonable for other occupants to let them know that the building should be evacuated. If a person with hearing difficulties is likely to be alone, then consider other means of raising the alarm. Among the most popular systems are visual beacons and vibrating devices or pagers that are linked to the existing fire alarm.

Voice alarms

Research has shown that some people do not always react quickly to a conventional fire alarm. Residents in your home may include people, such as children, who react badly to the noise of an alarm. You may therefore wish to consider using a voice alarm which can also incorporate a public address facility. The message or messages sent must be carefully considered. It is therefore essential to ensure that voice-alarm systems are designed and installed by a person with specialist knowledge of these systems.

Schematic plan

In order to quickly determine where a fire has been detected you should consider displaying a schematic plan, showing fire alarm zones in a multi-zoned system adjacent to the control panel.

2.1 Manual call points

Manual call points, often known as 'break-glass' call points, enable a person who discovers a fire to immediately raise the alarm and warn other people in the premises of the danger.

People leaving a building because of a fire will normally leave by the way they entered. Consequently, manual call points are normally positioned at exits and storey exits that people may reasonably be expected to use in case of fire, not just those designated as fire exits. You may wish to also locate some where they can be reached quickly by your staff. However, it is not necessary in every case to provide call points at every exit.

Manual call points should normally be positioned so that, after all fixtures and fittings, machinery and stock are in place, no one should have to travel more than 45m to the nearest alarm point. This distance may be less if your premises cater for people of limited mobility or there are particularly hazardous areas. They should be conspicuous (red), fitted at a height of about 1.4m (or less for premises with a significant number of wheelchair users), and not in an area likely to be obstructed.

2.2 Automatic fire detection

All residential care homes should have automatic fire detection, which actuates the fire warning system.

Automatic fire detectors can detect smoke, heat (or a combination of both) or carbon monoxide. Areas fitted with detection might include boiler rooms, laundries, kitchens, high risk and unoccupied areas, storage areas and walk in cupboards, large day rooms, access rooms to bedrooms, bedrooms, roof voids and areas or buildings served by a single stairway. As well as sounding the fire warning, the system should also operate any automatic door releases or smoke control and ventilation systems.

Smoke detectors are most appropriate as they provide the quickest response to any fire. However, the potential for false alarms must also be considered in the design of any system (see reducing false alarms below).

In small homes, e.g. with no more than one floor above ground, accommodating up to four residents who do not need assistance to escape, a basic system based on mains powered interlinked smoke alarms with battery back-up may provide suitable protection.

In larger premises, where all the residents are capable of evacuating themselves without staff assistance, the system should include detectors in all the escape routes, in any room opening onto any escape route and in any other areas of high risk. BS 5839-1[16] gives guidance on category L2 systems, which are designed to provide this level of protection.

In other larger premises, particularly those accommodating more than ten residents above the ground floor, or where a significant proportion of the residents are dependant upon staff assistance to escape in the event of a fire, a more comprehensive system of detection covering all areas of the building described above would be appropriate. BS 5839-1[16] gives guidance on category L1 systems which are designed to provide this higher level of protection.

2.3 Reducing false alarms

False alarms from automatic fire detection systems are a major problem and result in many unwanted calls to the fire and rescue service every year. Guidance on reducing false alarms has been published by ODPM/CFOA/BFPSA.[15]

If there are frequent false alarms in your premises, people may become complacent and not respond correctly to a warning of a real fire. In such circumstances, you may be failing to comply with fire safety law. All false alarms should be investigated to identify the cause of the problem and remedial action taken.

To help reduce the number of false alarms, the system design and location of detection and activation devices should be reviewed against the way the premises are currently used. For example, if a smoke detector is used in a bedroom the likelihood of false alarms caused by steam from an en-suite bath/shower room is increased. A heat detector may therefore be more suitable, provided the findings of a fire risk assessment show this to be more appropriate in each individual case. Similarly, if a manual call point is placed in an area where there is continual movement of equipment, the call point is more likely to be accidentally damaged. In this case a simple, fabricated hinged metal guard around the call point is likely to solve the problem.

Occasionally people set off a manual call point in the genuine, but incorrect, belief that there is a fire. Nothing should be done to discourage such actions and the number of false alarms generated this way is not significant.

Further detailed guidance on reducing false alarms is available in BS 5839-1.[16]

2.4 Staged fire alarms

In many premises sounding the fire warning system should trigger the immediate and total evacuation of the building. However, in some large or complex premises, this may not be necessary as alternative arrangements may be in place.

These alternative arrangements broadly fall into two groups. Firstly, those people potentially most at risk from a fire, usually those closest to where the alarm was activated, will be immediately evacuated, while others in the premises are given an alert signal and will only evacuate if it becomes necessary. This is generally called a phased evacuation and the initial movement, depending on the layout and configuration of the premises, can be either horizontal or vertical, but in residential care premises is most likely to be horizontal (see Part 1, Step 3.4.3 – Progressive horizontal evacuation).

The second alternative is for the initial alert signal to be given to certain staff, who then carry out pre-arranged actions to help others to evacuate more easily. It requires able, fully-trained staff to be available at all times and should not be seen as a simple means of reducing disruption to working practices. Where staged alarms are being used, disabled people should be alerted on the first stage to give them the maximum time to escape.

These arrangements both require fire warning systems capable of giving staged alarms, including an 'alert signal' and a different 'evacuate signal', and should only be considered after consultation with specialist installers and, if necessary, the relevant enforcing authority.

Such systems also require a greater degree of management input to ensure that staff and others are familiar with the system and action required.

2.5 Testing and maintenance

Your fire warning and/or detection system should be supervised by a competent person, given enough authority and training to manage all aspects of the routine testing and scrutiny of the system.

The control and indicating equipment should be checked at least every 24 hours to ensure there are no specific faults. All types of fire warning systems should be tested once a week. For electrical systems a manual call point should be activated (using a different call point for each successive test), usually by inserting a dedicated test key (see Figure 18). This will check that the control equipment is capable of receiving a signal and in turn, activating the warning alarms. Manual call points may be numbered to ensure they are sequentially tested.

Figure 18: Using a test key

It is good practice to test the alarm at the same time each week, but additional tests may be required to ensure that staff or people present at different times are given the opportunity to hear the alarm.

Where systems are connected to a central monitoring station, arrangements should be made prior to testing to avoid unwanted false alarms.

Six-monthly servicing and preventive maintenance should be carried out by a competent person with specialist knowledge of fire warning and automatic detection systems. This task is normally fulfilled by entering into a service contract with a specialist fire alarm company.

Testing of the system should be carried out by a competent person. Further guidance on testing and maintenance of fire warning systems can be found in British Standard 5839 Part 1.[16]

It is good practice to record all tests, false alarms and any maintenance carried out.

2.6 Guaranteed power supply

As residential care homes are likely to need an electrical fire warning system the Health and Safety (Safety Signs and Signals) Regulations 1996[5] requires it to have a back-up power supply.

Whatever back-up system is used, it should normally be capable of operating the fire detection and warning system for a minimum period of 24 hours and sounding the alarm signal in all areas for 30 minutes.

2.7 New and altered systems

Guidance on the design and installation of new systems and those undergoing substantial alterations is given in BS 5839.[16] If you are unsure that your existing system is adequate you will need to consult a competent person.

Section 3 Further guidance on firefighting equipment and facilities

You have responsibility for the provision of appropriate firefighting equipment. It is also your responsibility to check that all firefighting equipment is in the correct position and in satisfactory order before the premises are used.

Appropriate staff should be trained in the use of all such equipment.

3.1 Portable firefighting equipment

Fire extinguishers provided should be appropriate to the specific risks found in your premises in accordance with Table 1. This table also shows the different classes of fire, according to what is burning.

Table 1: Class of fire

Class of fire	Description
Class A	Fires involving solid materials such as wood, paper or textiles
Class B	Fires involving flammable liquids such as petrol, diesel or oils
Class C	Fires involving gases
Class D	Fires involving metals
Class F	Fires involving cooking oils such as deep-fat fryers

Note: If there is a possibility of a fire in your premises involving material in the shaded boxes then you should seek advice from a competent person.

Number and type of extinguishers
Typically for the Class A fire risk, the provision of at least one water-based extinguisher for approximately every 200m² of floor space, with a minimum of two extinguishers per floor, will normally be adequate. Depending on the outcome of your fire risk assessment, it may be possible to reduce this to one extinguisher in very small premises with a floor space of less than 90m².

Where it is determined that there are additionally other Classes of fire risk, the appropriate type, number and size of extinguisher should be provided. Further information is available in BS 5306-8.[18]

Where the fire risk is not confined to a particular location, e.g. Class A fires, the fire extinguishers should be positioned on escape routes, close to the exit from the room or floor, or the final exit from the building or, if necessary, adjacent. Similarly, where the particular fire risk is specifically located, e.g. flammable liquids, the appropriate fire extinguisher should be near to the hazard, so located that they can be safely used. They should be placed on a dedicated stand or hung on a wall at a convenient height so that employees can easily lift them off (at about 1m for larger extinguishers, 1.5m for smaller ones, to the level of the handle). Ideally no one should have to travel more than 30m to reach a fire extinguisher.

Where there are self-contained small premises within the larger premises, multi-purpose extinguishers which can cover a range of risks may be appropriate.

Consider the implications of the Manual Handling Operations Regulations 1992[17] when selecting and siting firefighting equipment.

In some care homes your fire risk assessment may need to take account of the possibility of an extinguisher being used as a weapon or being vandalised and therefore they may need to be located in positions where staff have control.

Extinguishers manufactured to current Standards (BS EN 3-7[85]) are predominantly red but may have a colour-coded area, sited above or within the instructions, denoting the type of extinguisher. Most older extinguishers, manufactured to previous standards, have bodies painted entirely in a single colour which denotes the type of extinguisher. These older extinguishers remain acceptable until they are no longer serviceable. However, it is good practice to ensure that old and new style extinguishers are not mixed on the same floor of a building.

The following paragraphs describe the different types of extinguisher. The colour referred to is the colour of the extinguisher or the colour-coded area.

Water extinguishers (Red)
This type of extinguisher can only be used on Class A fires. They allow the user to direct water onto a fire from a considerable distance. A 9-litre water extinguisher can be quite heavy and some water extinguishers with additives can achieve the same rating, although they are smaller and therefore considerably lighter. This type of extinguisher is not suitable for use on live electrical equipment.

Water extinguishers with additives (Red)
This type of extinguisher is suitable for Class A fires. They can also be suitable for use on Class B fires and where appropriate, this will be indicated on the extinguisher. They are generally more efficient than conventional water extinguishers.

Foam extinguishers (Cream)
This type of extinguisher can be used on Class A or B fires and is particularly suited to extinguishing liquid fires such as petrol and diesel. They should not be used on free-flowing liquid fires unless the operator has been specially trained, as these have the potential to rapidly spread the fire to adjacent material. This type of extinguisher is not suitable for deep-fat fryers or chip pans.

Powder extinguishers (Blue)
This type of extinguisher can be used on most classes of fire and achieve a good 'knock down' of the fire. They can be used on fires involving electrical equipment but will almost certainly render that equipment useless. Because they do not cool the fire appreciably it can re-ignite. Powder extinguishers can create a loss of visibility and may affect people who have breathing problems and are not generally suitable for confined spaces.

Carbon dioxide extinguishers (Black)
This type of extinguisher is particularly suitable for fires involving electrical equipment as they will extinguish a fire without causing any further damage (except in the case of some electrical equipment, e.g. computers). As with all fires involving electrical equipment, the power should be disconnected if possible.

Class 'F' extinguishers
This type of extinguisher is mostly suitable for large kitchens with deep-fat fryers.

Selection, installation and maintenance of portable fire extinguishers

All portable fire extinguishers will require periodic inspection, maintenance and testing. Depending on local conditions such as the likelihood of vandalism or the environment where extinguishers are located, you will need to carry out brief checks to ensure that they remain serviceable. In normal conditions a monthly check should be enough. Maintenance by a competent person should be carried out annually.

New fire extinguishers should comply with BS EN 3-7.[85] Guidance on the selection and installation of fire extinguishers is given in BS 5306-8,[18] for maintenance in BS 5306-3[19] and for colour coding in BS 7863.[20]

Fire blankets

Fire blankets should be located in the vicinity of the fire hazard they are to used on. They are particularly suitable for kitchens. They need to be in a position that can be safely accessed in the event of a fire. They are classified as either light-duty or heavy-duty. Light-duty fire blankets are suitable for dealing with small fires in residential care premises involving cooking oils, fat and clothing.

3.2 Fixed firefighting installations

These are firefighting systems which are normally installed within the structure of the building. They may already be provided in your premises or you may be considering them as a means of protecting some particularly dangerous or risk-critical area as part of your risk-reduction strategy.

Hose reels

Permanent hose reels (see Figure 19) installed in accordance with BS EN 671-3[21] provide an effective firefighting facility. They may offer an alternative, or be in addition to, portable firefighting equipment. A concern is that untrained people will stay and fight a fire when escape is the safest option. Where hose reels are installed, and your fire risk assessment expects relevant staff to use them in the initial stages of a fire, they should receive appropriate training.

Maintenance of hose reels includes visual checks for leaks and obvious damage should be carried out regularly and more formal maintenance checks should be carried out at least annually by a competent person.

Figure 19: Hose reel

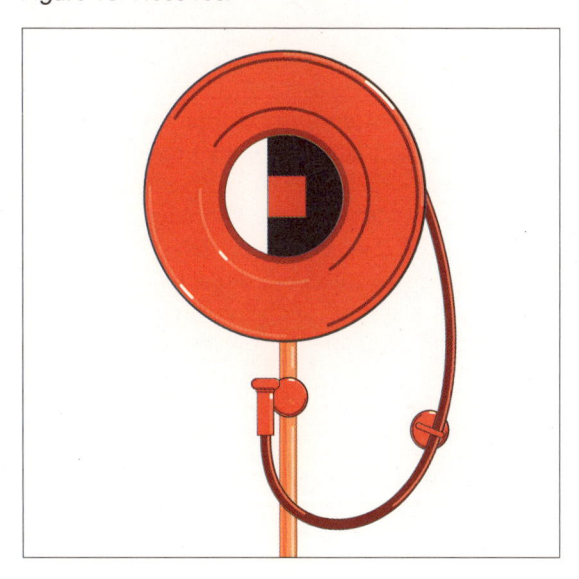

Sprinkler systems

Sprinkler systems can be very effective in controlling fires. They can be designed to protect life and/or property and may be regarded as a cost-effective solution for reducing the risks created by fire. Recent research by BRE[74] has shown that for residential care premises, in particular, there can be worthwhile safety and financial benefits in fitting a sprinkler system.

Where installed, a sprinkler system is usually part of a package of fire precautions in a building.

Sprinkler protection can give additional benefits, such as a reduction in the amount of portable fire-fighting equipment necessary or a relaxation of restrictions in the design of buildings (including the lengths of escape routes, the provision of fire-resisting barriers or the provision of some self-closing fire doors). There may also be additional benefits such aas a reduction in insurance premiums and a significant reduction in the risks of major disruption to your activities as a care provider.

If you have, or intend to install, a sprinkler system, you should seek specialist advice on the esigna dn installation of the system, and also on the benefits it can provide in respect of your risk assessment.

Guidance on the design and installation of new sprinkler systems and the maintenance of all systems is given in BS EN 12845,[22] the Loss Prevention Council (LPC) Rules,[88] BS 5306-2[89] or BS 9251.[87]

Routine maintenance by on-site competent people may include checking of pressure gauges, alarm systems, water supplies, any anti-freezing devices and automatic booster pump(s).

A competent maintenance contractor should carry out all major servicing and can provide guidance on what records need to be completed.

If a sprinkler system forms an integral part of your fire strategy it is imperative that adequate management procedures are in place to cater for those periods when the sprinkler system is not functional. These should form part of your emergency plan. Although the actual procedures will vary, such measures may include the following:

- Restore the system to full working order as soon as possible.
- Limit any planned shutdown to low-risk periods when numbers of people are at a minimum or when the building is not in use. This is particularly important when sprinklers are installed to a life safety standard or form part of the fire safety engineering requirements.
- Avoid higher-risk processes such as 'hot-work'.
- Training staff in extra duties and conducting fire patrols.
- Revising any evacuation strategy.
- Carrying out maintenance on a zoned basis, to prevent the whole system being out of commission at the same time; and
- Inform the local fire and rescue service.

Other fixed installations

There are a number of other fixed installations including water mist, gaseous, deluge and fixed powder systems. If your premises have a fixed firefighting system that you are unfamiliar with, then seek advice. Where a fixed firefighting system forms an integral part of your fire safety strategy, it should be maintained by a competent person in accordance with the relevant standard.

3.3 Other facilities (including those for firefighters)

Building Regulations and other Acts, including local Acts, may have required firefighting equipment and other facilities to be provided for the safety of people in the building and to help firefighters. Fire safety law places a duty on you to maintain such facilities in good working order and at all times.

These may include:

- access roads for fire engines and firefighters;
- firefighting shafts and lifts;
- fire suppression systems, e.g. sprinklers, water mist and gaseous;
- smoke-control systems;
- dry or wet rising mains and firefighting inlets;
- information and communication arrangements e.g. fire telephones and wireless signals; and
- firefighters' switches.

The Workplace (Health, Safety and Welfare) Regulations 1992[23] also require that systems provided for safety within a workplace are maintained.

You must arrange any necessary contacts with the emergency services and it may be appropriate to invite the fire and rescue service to familiarise themselves on layouts and fire systems as a precautionary measure.

Access for fire engines and firefighters

Buildings that have been constructed to modern Building Regulations or in accordance with certain local Acts will have been provided with facilities that allow fire engines to approach and park within a reasonable distance so that firefighters can use their equipment without too much difficulty.

These facilities may consist of access roads to the building, hard standing areas for fire engines and access into the building for firefighters. It is essential that where such facilities are provided they are properly maintained and available for use at all times.

Where a building or complex is used by a number of different occupants you will need to ensure co-operation between the various responsible people to maintain fire and rescue service access. In exceptional cases, where access is persistently obstructed, you may need to make additional arrangements.

See Approved Document B to the Building Regulations[24] for more information.

Other firefighting facilities

You may have other facilities to assist firefighters installed in your premises in addition to these listed above. These should be properly maintained by a competent person and included in your maintenance audit (see Appendix A for an example checklist). Such facilities can include:

- information signs for firefighters;
- static water supplies, private hydrants, meter bypass valves and underground tanks; and
- standby fire pumps, electrical generators, air pumps and hydraulic motors.

Section 4 Further guidance on escape routes

Introduction

This section provides further guidance on the general principles that apply to escape routes and provides examples of typical escape route solutions for a range of building layouts. The guidance is based on premises of normal risk so, if your premises (or parts of your premises) are higher (or lower) risk you should adapt the solution accordingly.

You are not obliged to adopt any particular solution for escape routes in this section if you prefer to meet the relevant requirement in some other way. If you decide to adopt some alternative arrangement it will need to achieve at least an equivalent level of fire safety.

Refer to the glossary (Appendix D) for the definitions of any terms you may not be familiar with.

Levels of risk

In order to apply the guidance in this section, you need to understand that in any fire situation, the time that people have to escape before they could become affected by the fire is limited. Providing them with sufficient time usually means that as well as having an appropriate way of detecting and giving warning in case of fire, the distance that people have to travel to make their escape to a place of reasonable or total safety must be restricted.

The travel distances which are usually appropriate for this purpose (and are suggested later in this section) vary according to the level of risk in the premises (or part of them). To check your escape routes you will need to form a judgement about the level of risk that people may be at after you have taken other risk reduction (preventative and protective) measures.

In premises where there is a likelihood of a fire starting and spreading quickly (or a fire could start and grow without being quickly detected and a warning given) and affect the escape routes before people are able to use them then the risk should normally be regarded at "higher". Such premises could include those where significant quantities of flammable materials are used or stored; ready sources of ignition are present, premises where significant numbers of the people present are likely be unable to move without assistance; premises where the construction provides hidden voids or flues through which a fire could quickly spread and premises where the layout prohibits or constrains provision of the sub compartments required for progressive horizontal evacuation (see page 68).

In premises where there is a low occupancy level and all the occupants are able bodied and capable of using the means of escape without assistance; very little chance of a fire; few if any highly combustible or flammable materials or other fuels for a fire; fire cannot spread quickly; and will be quickly detected so people will quickly know that a fire has occurred and can make their escape, then the risk can usually be regarded as 'lower'.

In most cases however, the risk will usually be normal.

The travel distances suggested are not hard and fast rules and should be applied with a degree of flexibility according to the circumstances. For example, in premises where the risk might otherwise be considered 'normal' but where there are a significant number of people who move slowly or may need assistance to evacuate, it would usually be appropriate to consider this a 'higher' risk. However, where other measures are in place to mitigate this, such as the availability of extra assistance and this has been planned for in your emergency plan, it may be that the risk level can be regarded as 'normal to higher'.

Equally, in premises where the risk category would otherwise be 'lower' but for the fact that a small number of occupants may move slowly or need assistance, it may be appropriate to categorise the risk as 'normal' in these circumstances.

If you are not sure about the level of risk that remains in your premises, you should seek advice from a competent person.

4.1 General principles

Evacuation strategies

Single stage evacuation

In the event of a fire, it is normally preferable to evacuate people as quickly as possible from the premises. Escape routes in a building should be designed so that people can escape quickly enough to ensure they are not placed in any danger from fire. The time available will depend on a number of factors, including how quickly the fire is detected and the alarm raised, the number of escape routes available, the nature of the occupants and the speed of fire growth. For simplicity, the travel distances in Part 2, Table 2 on page 76 takes these factors into account. Part 2, Section 4 on page 67 will help you decide the level of risk in your premises for escape purposes.

At the design stage of a building, the lengths of escape routes are kept within certain guidelines so that in the event of a fire, the occupants of the premises can escape in the shortest possible time.

This strategy is appropriate for many types of care home, where it may be expected that all people inside are able to (and will) evacuate quickly to outside the building to a place of total safety. This strategy will also usually be appropriate in more complex premises for visitors, wakeful mobile residents in day rooms and ancillary staff not required to assist with evacuation of other residents.

Progressive horizontal evacuation

Other than small premises where residents can escape quickly and immediately to a place of total safety in open air (single stage evacuation), all other care premises to which this guide applies should develop a strategy of progressive horizontal evacuation.

The process of horizontal evacuation relies on the parts of the building that are used for sleeping accommodation being separated into smaller sub-compartments called protected areas. These are areas separated from each other on the same level by walls and doors that provide at least 30 minutes of fire resistance. Each protected area should have at least two alternative exits to adjoining areas.

If a fire occurs, residents can be moved away from the affected area to the adjoining protected area, where they are able to remain for a time in reasonable safety.

Protected areas should be designed to provide:

- Sufficient capacity to accommodate the number of occupants who will need to use them. For this purpose a protected area should be sufficient capacity to accommodate its normal occupants and the occupants of the largest adjoining protected area.

- Progressive movement away from a fire via sequential adjoining protected areas.

- Means for escape via stairway(s) should this become necessary.

The number and size of the protected areas depends on a number of factors:

- the time it will take to evacuate people from the area of a fire to an adjacent protected area (see below);
- the number of people to be evacuated;
- the level of any mobility impairment;
- the number of staff to assist in evacuation;
- the fire protection arrangements;
- layout of the premises; and
- location and number of staircases;

Each of the above factors will also influence the time needed to move the occupants from the fire area into the reasonable safety of an adjoining protected area. A simple assessment can be made to determine the time more accurately by conducting a simulated evacuation exercise, involving residents where possible, or volunteers acting as residents. A 'worst case scenario' should always be assumed to give an accurate reflection of the longest time it would take to evacuate. This is usually when the lowest number of staff are available (e.g. at night when residents are asleep).

Figure 21 illustrates a typical protected area arrangement.

Figure 21: Protected areas in horizontal evacuation

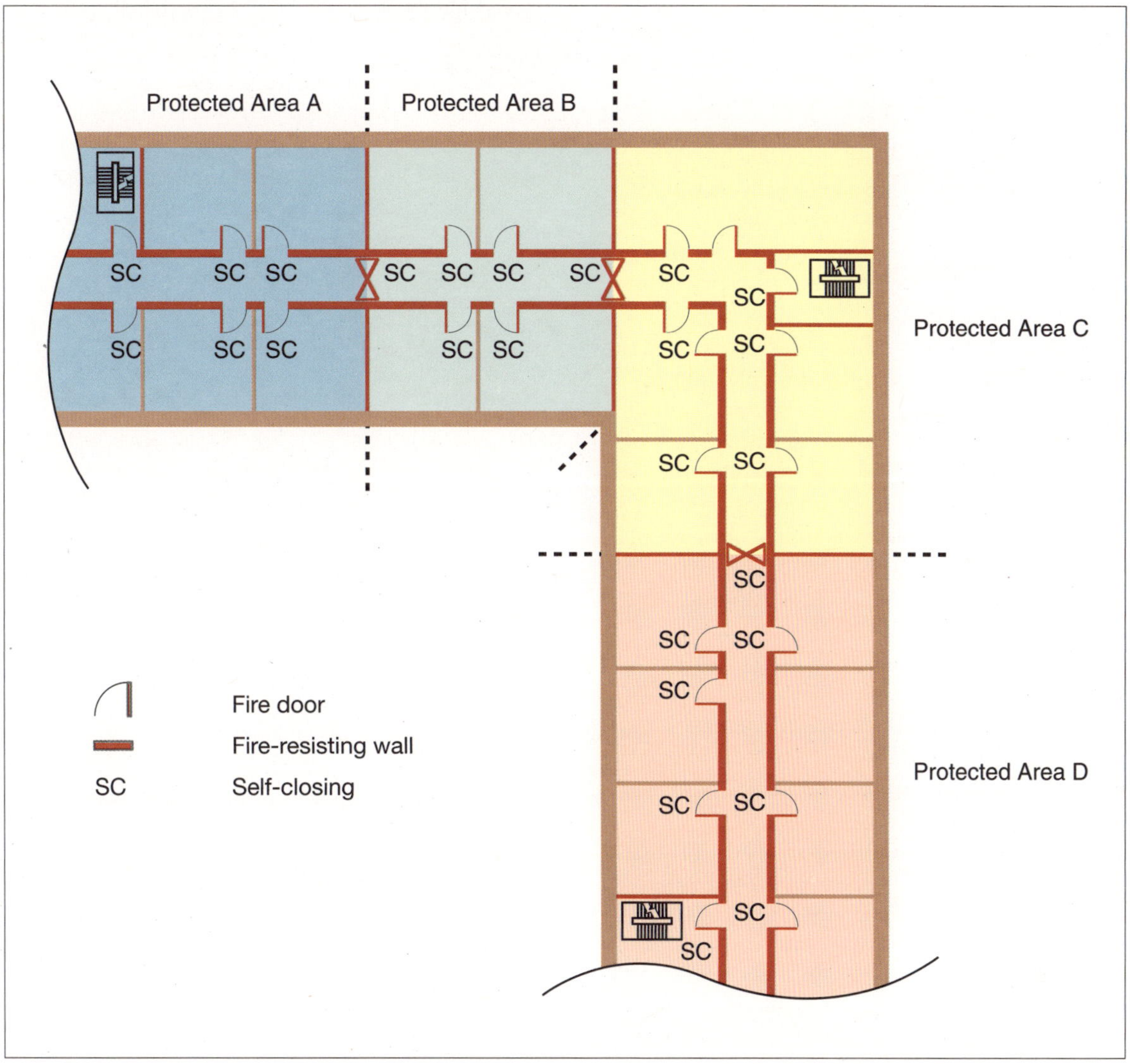

Evacuation times from protected areas

The evacuation time from a protected area to an adjacent protected area or other place of reasonable or total safety may be considered in 3 stages:

- The risk of the protected area should be determined (high, normal or low) refer to Section 4, page 67.

- The dimension of the protected area should be such that the travel distances in Table 2, page 76 can be met (by refering to the appropriate risk level).

- It should be possible under normal operating conditions to evacuate any given protected area in 2½ minutes using the staff present to assist residents, where necessary.

Studies of human behaviour in an emergency situation have shown that about two thirds of the time available to escape is taken up by the initial reaction to the developing situation. For example, people will decide whether the situation is real or false, often waiting to see the reaction of people around them and generally gathering information to decide whether to act or not. The final third is taken up by the actual movement away from the area of fire.

You should aim to evacuate all occupants from a protected area (sub compartment) to a place of reasonable or total safety within 2½ minutes of the alarm being raised. Staff training and pre-planning should assist to minimise the time to react to the alarm so as to maximise the time available to assist residents to evacuate.

To assist evacuation consider accommodating residents who can make their own way to safety unaided on the upper levels, with people who have mobility impairment closer to the ground floor. As mobility increases it may be possible to accommodate more residents in any one protected area.

If you cannot achieve this you need to estimate how fast you expect people to move, either by their own efforts or assisted by staff, so that you can determine the size and location of each protected area and the number of beds that can be accommodated safely and make any necessary adjustments to the protected areas and/or your emergency plan.

For non-ambulant or semi-ambulant residents, who cannot make their way to a place of safety unaided, there needs to be adequate and protected escape routes to permit the staff to reach the residents.

Issues to consider include:

- Where there is occupancy of mixed mobility it will be safer to base the evacuation speed on the slowest moving category.

- It is essential that there are sufficient staff available at all times to implement your emergency plan, and that staff are trained to react promptly to the fire alarm, in order to maximise the time available for evacuation.

- Feature such as stairs, ramps and doors across corridors will also have a bearing on the evacuation speed of each category.

- When carrying out a review you should re-evaluate the escape time as peoples' mobility will change over a period of time.

- Since residents who need assistance can only be moved as and when carers become available, it follows that (depending upon the numbers of staff) these residents may not be moving to safety concurrently.

In situations where residents of poor mobility are to be physically moved by staff, you must ensure that the staff have received the appropriate training to do so. If evacuation aids or equipment is provided for this purpose, staff should be trained to deploy and utilise them as quickly and efficiently with as little distress to the resident as possible. Determination of acceptable escape times depends on all of the above, and no single factor should be considered in isolation. The ability of staff to conduct an evacuation of each protected area within 2½ minutes of the alarm being raised is a starting point upon which to make as assessment. Extended escape time may be able to be dealt with by careful adjustment to the above factors.

You should conduct a simulated evacuation of each protected area and assess whether the evacuation times achieved are reasonable and consistent with those recommended for the levels of mobility likely to be found. If you conclude that movement times to an adjoining protected area are excessive then the following options should be considered:

- reducing the size of the protected area, (this may be no more than dividing the wing of a building by positioning a set of fire-resisting self closing doors across a corridor);

- providing additional staff to assist with the horizontal evacuation;

- accommodating less people in the protected area, or

- providing additional fire protection systems, such as an appropriately designed sprinkler system, to the protected areas to suppress the fire growth.

Alternatively, fire-engineering solutions may be available to compensate for extended escape time, but should only be considered following the advice of an expert.

Delayed evacuation
Exceptionally, in some care homes it may be difficult or dangerous (because of medical conditions or treatments) to get all semi-ambulant and non-ambulant residents into an adjoining protected area, or to a refuge during the initial stage of the evacuation.

In this situation, the individual bedroom(s) may need to provide a temporary refuge to protect the occupant from a fire elsewhere in the building until they can be taken to a place of total safety, or the danger has passed. This can be done by enclosing the bedroom in an enhanced level of fire-resisting construction (protected bedrooms).

A protected bedroom should be of 60 minute fire-resisting construction and the door should be fire-resisting and fitted with a self-closing device. In addition the escape route from the protected bedroom(s) to the adjoining protected areas, refuge or final exit (including any stairway) will also require an increased level of fire protection to allow access for staff to assist with subsequent evacuation from the protected bedroom(s).

If necessary the door may be fitted with electromechanical hold-open or free swing devices that operate immediately the fire alarm actuates.

If provision of such fire resistance is not possible, you may be able to show through your risk assessment that alternative measures to limit the growth and spread of the fire are appropriate, such as an automatic fire suppression system supported by robust staff response procedures.

Any resident who is initially left in a fire protected bedroom should be accompanied by a carer. As such, the total number of residents awaiting evacuation in protected bedrooms should be less than the number of staff on duty. It is imperative that if some less able residents are left in protected bedrooms to await evacuation, then other staff know which rooms have been evacuated and those which still contain residents and where necessary are able to notify the fire and rescue service when they arrive.

Arrangements for delayed evacuation should only be based on a pre-planned basis.

Communication procedures should let residents, who can hear an alarm but not able to respond, know what is happening.

Whichever strategy or combination of strategies you adopt it is likely to be influenced by a number of factors including:

- size, type, layout and construction of the premises;
- time of the day;
- reaction, mobility and dependency of residents;
- number of staff; and
- nature and location of occupants, e.g. visitors, residents, ancillary staff etc.

It is essential that your evacuation strategy is fully detailed in your emergency plan and included in your staff training programme (see Section 7).

Suitability of escape routes
You should ensure that your escape routes are:

- suitable;
- easily, safely and immediately usable at all relevant times;
- adequate for the number of people likely to use them;
- usable without passing through doors requiring a key or code to unlock;
- free from any obstructions, slip or trip hazards;
- well lit by normal or emergency escape lighting; and
- available for access by the emergency services.

Ideally all doors on escape routes should open in the direction of escape. This is particularly important if more than 60 people use them or they provide an exit from an area of high fire risk.

While not normally acceptable, the use of ladders, floor hatches, wall hatches or window exits may be suitable for small numbers of able-bodied, trained staff in exceptional circumstances, e.g. for maintenance staff accessing plant.

Number and type of people using the premises
A fire occurring in a residential care establishment that results in the need for any evacuation will usually involve members of staff to assist in the movement of residents away from the area of the fire to an area of safety. How this can be achieved will depend, to a large degree, on

the mobility of the residents, how well they respond to the emergency situation and the level of co-operation they are able to give.

At night, this situation will be made more difficult by the fact that most residents will be sleeping, with the responses of some being further impaired by medication. In many care premises it will not be practicable to carry out an immediate full evacuation and in any case this may be harmful to the residents. Therefore, the most appropriate method of initial evacuation is to adopt a policy of horizontal movement to a place of reasonable safety within the building.

As your escape routes need to be adequate for the people likely to use them you will need to consider how many people, including staff, residents and visitors, may be present at any one time. Where premises have been subject to building regulations approval for use providing residential care, the number and width of escape routes and exits will normally be enough for the anticipated number of people using the building. In such buildings and in particular where the risk has changed or buildings were constructed before national building regulations it is still necessary to confirm the provision as part of your risk assessment.

For most care premises, the maximum numbers of staff, residents, visitors and contractors liable to be in the building at the same time will be known by the responsible person who will normally be aware from a personal knowledge of visiting patterns.

If you propose to make changes to the use or layout of the building which may increase the number of people, you will need to check the design capacity by referring to guidance given in the Building Regulations Approved Document B.[24] and consult the Building Control body.

Staff numbers
The numbers of staff in your premises will have a very significant influence on the outcome of your risk assessment.

You may need to consult your registration authority to determine or agree the numbers of staff you should have in attendance, in particular at night.

In most residential care premises, staff are always present and are expected to play a role in evacuation. Should a fire start it will be first necessary to evacuate the sub-compartment of origin, and the speed of evacuation will be influenced by the number of staff available.

The speed of evacuation and the number of residents who can be evacuated before staff are exhausted will depend upon the number of staff available, the immediacy of staff availability, the distance staff have to travel to commence the evacuation and the physical responsiveness of the residents to being evacuated. Therefore the number of resident beds which you allow in each sub-compartment should depend on the minimum number of your staff who are awake and available on the premises (normally the night-time staffing level). Your risk assessment should identify the number of staff you need to carry out your emergency plan. It follows that where there is a need for staff intervention to carry out your emergency plan, then staff numbers at all times must be sufficient; you should be able to demonstrate to the enforcing authority that you always have sufficient staff to effectively carry out your emergency plan without the assistance of outside agencies (e.g. by a fire drill).

Mobility impairment and level of awareness of residents.
Some mentally ill or elderly residents' level of understanding during an emergency may be such that they may resist the assistance from staff due to fear and confusion. It may take time to calm them down sufficiently for evacuation to take place with their co-operation. Sedation due to medication may impede evacuation as more members of staff may be required to move persons with slow responses.
In parts of care homes where residents have such impediments the options to ensure quick evacuation are likely to involve either reducing the size of the protected area or increasing the number of staff.

The degree to which residents can self evacuate or need support will vary considerably depending on their mental health or the progression of their condition. Staffing requirements are likely to change over time as the dependency levels of clients change.

People with minor mobility problems can often use the normal means of escape, albeit with some difficulty, but those people unable to walk will probably have to rely on assistance from staff.

Fire-resisting construction

The type and age of construction are crucial factors to consider when assessing the adequacy of the existing escape routes. To ensure the safety of people it will be necessary to protect escape routes from the effects of a fire. In older premises (see Appendix C for more information on historical properties) it is possible that the type of construction and materials used may not perform to current fire standards. Also changes of occupier and refurbishment may have led to:

- cavities and voids being created, allowing the potential for a fire to spread unseen;
- doors worn by age and movement being less likely to limit the spread of smoke;
- damaged or lack of cavity barriers in modular construction; and
- breaches in fire compartment walls, floors and ceilings created by the installation of new services, e.g. computer cabling.

All compartment floors in care premises should be of fire resisting construction. In general a rating of 30 minutes fire resistance is acceptable however 60 minute rating is appropriate where delayed evacuation is necessary (see delayed evacuation on page 71).

Where an escape route needs to be separated from the rest of the premises by fire-resisting construction, e.g. a dead-end corridor or protected stairway (see Figures 29 and 34), then you should ensure the following:

- Doors (including access hatches to cupboards, ducts and vertical shafts linking floors) walls, floors and ceilings protecting escape routes should be capable of resisting the passage of smoke and fire for long enough so that people can escape from the building;
- Where suspended or false ceilings are provided, the fire resistance should extend up to the floor slab level above. For means of escape purposes a 30 minutes fire-resisting rating is normally enough; and
- Cavity barriers, fire stopping and dampers are appropriately installed.

It may be necessary to upgrade the standard of construction of the floors, walls and ceilings to provide 30-minute fire-resisting construction to ensure the occupants of the premises can be evacuated to a place of reasonable safety within a safe time. This will be crucial when considering providing further protected areas.

Except where otherwise demonstrated by your fire risk assessment, each bedroom should be seperated from an adjoining hall/corridor by a 30-minute fire-resisting construction. Doors should be of a similar fire-resisting standard, and be self-closing.

The ability of residents to move freely within the premises should not be affected by any fire safety provision. There may therefore be a need to use electromagnetic hold-open or free-swing devices on appropriate doors.

If your premises have been previously registered for use as a residential care facility, then it is likely that the structure has been assessed and is considered satisfactory for horizontal evacuation. Any removal or alteration of the structure may have a serious impact on the safety of residents and should only done where a comprehensive assessment and consultation has been carried out.

If there is any doubt about the nature of the construction of your premises, ask for advice from a competent person.

Number of escape routes and exits

At the building design or proposed alteration stage, the anticipated length of the escape routes is kept within certain acceptable limits. In the case of residential care premises it may not be the length of the escape routes, but the time taken for all the residents at risk to be evacuated to an adjoining area of reasonable safety. However, there will be other occupants of the premises, such as visitors and contractors etc., who will rely on conventional escape routes and procedures to escape from the premises.

In general there should normally be at least two escape routes from all parts of the premises. A route providing escape in one direction only (called dead ends) can be acceptable provided travel distances are within those recommended in Table 2.

Widths and capacity of escape routes and stairways

Once you have established the maximum number of people likely to be in any part of the premises, the next step is to establish that the capacity of the escape routes is adequate for people to escape safely in sufficient time to ensure their safety in case of fire.

At least two exits should be provided if a room/area is to be occupied by more than 60 persons. This criterion of 60 can be varied in proportion to the risk, for a lower risk there can be a slight increase, for a higher risk, lower numbers of persons should be allowed.

The effective usable width of an escape route is the narrowest point, normally a door or other restriction such as narrowing of a corridor due to fixtures and fittings. The effective width of a doorway is the clear unobstructed width through the doorway when the door is open at right angles to the frame. The effective width at any other point is the narrowest clear unobstructed width through which people can pass.

The minimum width of an escape route in your premises should ideally be 1,050mm but in any case not less than 750mm (unless it is for use by less than 5 people in part of your premises) and, 900mm where wheelchair users are likely to use it. Wider escape routes will be needed if you need to evacuate people in beds.

The aggregate width of all the escape routes should be not less than that required to accommodate the maximum numbers of people likely to use them.

When calculating the overall available escape route capacity for premises that have more than one way out, you should normally assume that the widest is not available because it has been compromised by fire. If doors or other exits leading to escape routes are too close to one another you should consider whether the fire could affect both at the same time. If that is the case, it may be necessary to discount them both from your calculation.

As a general rule stairways should be at least 1,050mm wide and in any case not less than the width of the escape routes that lead to them. In all cases the aggregate capacity of the stairways should be sufficient for the number of people likely to have to use them in case of fire.

Stairways wider than 2,100mm should normally be divided into sections, each separated from the adjacent section by a handrail, so that each section measured between the handrails is not less than 1,050mm wide.

Facilities for those with mobility impairment
Effective management arrangements need to be put in place for those that need help to escape.

If your escape plan requires disabled people to move horizontally to a refuge you should consider the following points:

- A refuge is a place of reasonable safety in which a disabled person can wait either for an evacuation lift or for assistance up or down stairs. Disabled people should not be left alone in a refuge area whist waiting for assistance with evacuation from the building. Depending on the design and fire resistance of other elements, a refuge could be a lobby, corridor, part of a public area or stairway, or an open space such as a balcony or similar place which is sufficiently protected (or remote) from any fire risk and provided with its own means of escape (see Figure 20).

- Where refuges are provided, they should be enclosed in a fire-resisting structure which leads directly to a protected escape route and should only be used in conjunction with effective management rescue arrangements. Your fire safety strategy should not rely on the fire and rescue service rescuing people waiting in these refuges.

- If firefighting lifts (provided in high buildings as firefighting access) are to be used for evacuation, this should be co-ordinated with the fire and rescue service as part of the pre-planned evacuation procedures.

- Normal lifts may be considered suitable for fire evacuation purposes if they have fire protection. They should be subject to an adequate fire risk assessment and development of a suitable fire safety strategy by a competent person.

- Since evacuation lifts can fail, a disabled person should having reached a refuge also be able to gain access to a stairway (should conditions in the refuge become untenable). An evacuation lift with its associated refuge should therefore be located adjacent to a protected stairway.

- There should be a means of communicating between the refuge and the person responsible for the evacuation.

- Enough escape routes should always be available for use by disabled people. This does not mean that every exit will need to be adapted. Staff should be aware of routes suitable for disabled people so that they can direct and help people accordingly.

- Where ramps are necessary for the emergency evacuation of people in wheelchairs they should be constructed in accordance with the guidance Approved Document M.[80]

In some premises it may be appropriate to accommodate people with a mobility or sensory impairment on the ground floor.

Further guidance on facilities for those with mobility impairment is available in BS 5588-8[81] and BS 5588-12.[75]

Assistance on stairways
Regardless of the evacuation strategy, if your escape plan involves the manual movement of mobility-impaired people down stairways there are a number of issues you should consider:

- stairways used for the emergency evacuation of disabled people should comply with the requirements for internal stairs in the building regulations. Specialist evacuation chairs or other equipment may be necessary to negotiate stairs;

- plans should allow for the careful carrying of disabled people down stairs without their wheelchairs, should the wheelchair be too large or heavy. You will need to take into account health and safety manual handling procedures in addition to the dignity and confidence of the disabled person;

- staff will always need to be available, adequately trained, able and willing to assist;

- the process will need to be regularly practised;

- movement down the stairs must not impede the escape of others;

- equipment appropriate for all people in the premises, such as emergency evacuation chairs, should be considered;

- disabled people should be consulted; and

- stairlifts should not be used for emergency evacuation. Where installed in a stairway used for emergency evacuation, no parts of the lift, such as its carriage rail, should be allowed to reduce the effective width of the stairway or any other part of an emergency evacuation route.

Figure 20: Example of a refuge

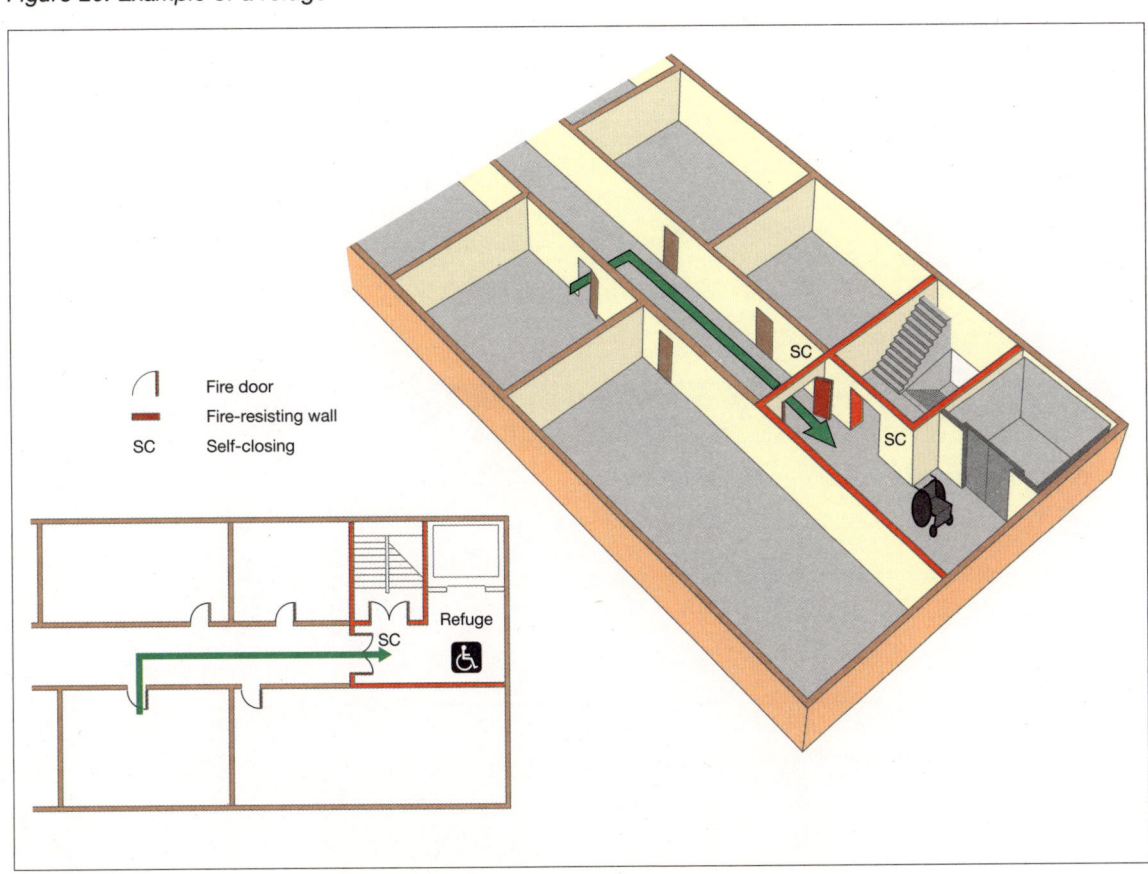

Travel distance

Having established the escape strategy to be adopted in your premises, you now need to confirm that the number and location of existing exits is adequate. This is normally determined by the distance people have to reach them.

In new buildings which have been designed and constructed in accordance with modern building standards the travel distances will already have been calculated. Once you have completed your fire risk assessment you need to confirm that those distances are still relevant.

Table 2 gives guidance on travel distances. It should be understood, however, that these distances are flexible and may be increased or decreased depending upon the level of risk after you have put in place the appropriate fire-prevention measures (Part 1, Step 3.3).

When assessing travel distances you need to consider the distance to be travelled by people when escaping allowing for walking around furniture etc. The distance should be measured from all parts of the premises to the nearest place of reasonable safety which is:

- a protected stairway (a storey exit);
- a separate protected area (fire compartment) from which there is a final exit to a place of total safety or a storey exit; or
- the nearest available final exit.

The distance to be travelled should be kept to as short as possible as increased distances will inevitably mean extended escape times. Introducing more protected areas within the building may reduce travel distances, but this may seriously impact on the usability of the premises by residents and should only be considered after other options have been explored.

The distance for initial horizontal movement suggested in Table 2 above should always be based on the assessment of needs and mobility of the residents.

Residential care homes that have been built to Building Regulations since 1995 will usually have been provided with escape routes and maximum distances of travel in accordance with those set out in Table 2.

Your current escape route tavel distances may be different from these since they may be based on recommendations made in alternative guidance.

Although travel distances are suggested in Table 2, it must be emphasised that the time taken to move residents will be the important factor in any fire situation.

Escape routes that start in normal risk areas and pass through higher risk areas should be avoided. Where this arrangement is unavoidable, the travel distances should be appropriate to the higher risk.

The suggested travel distances may be increased by the addition of further fire protection measures, e.g. automatic fire suppression, but decreased to allow for staff to reach non- or semi- ambulant residents.

Table 2 Suggested travel distances

Escape routes	Suggested range of travel distance
Where more than one escape route is provided	10m in higher fire-risk area[Note 1] 18m in normal fire-risk area 25m in lower fire-risk area[Note 2]
Where only a single escape route is provided	6m in higher fire-risk area[Note 1] 9m in normal fire-risk area 15m in lower fire-risk area[Note 2]

Note 1: Where there are small high-risk areas this travel distance should apply. Where the risk assessment indicates that the whole building is high risk, ask advice from a competent person.

Note 2: The travel distance for lower risk premises should only be applied in exceptional cases in the very lowest risk premises

Measuring travel distance
The figures that follow are schematic only and are intended to represent part of a larger building.

The route taken through the room or space will be determined by the layout of the contents (Figure 22). It is good practice to ensure the routes to the room exits are kept as direct and as short as possible, especially in accommodation where sleeping will occur, thus reducing the time taken to exit the room. In a small room, e.g. a bedroom, there will usually only be one exit, but in a larger room, e.g. a common room, there may be many exits.

In some cases where furniture is often moved around or the use is liable to frequent change, such as ballrooms or conference halls, you should ensure that exits from the room do not become inadvertently blocked or the escape route from the room is significantly extended.

Figure 22: Measuring travel distance

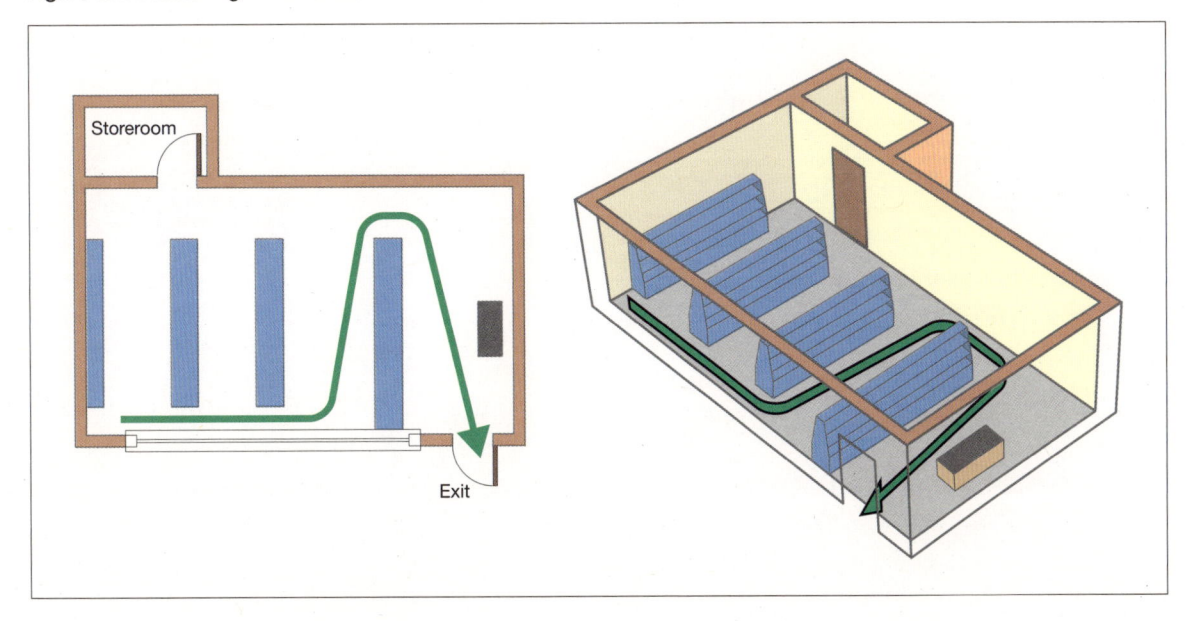

Alternative exits

Where alternative exits from a space or room are necessary they should wherever possible be located at least 45° apart (see Figure 23) unless the routes to them are separated by fire-resisting construction (see Figure 24). If in doubt consult a competent person.

Figure 23: Alternative exits

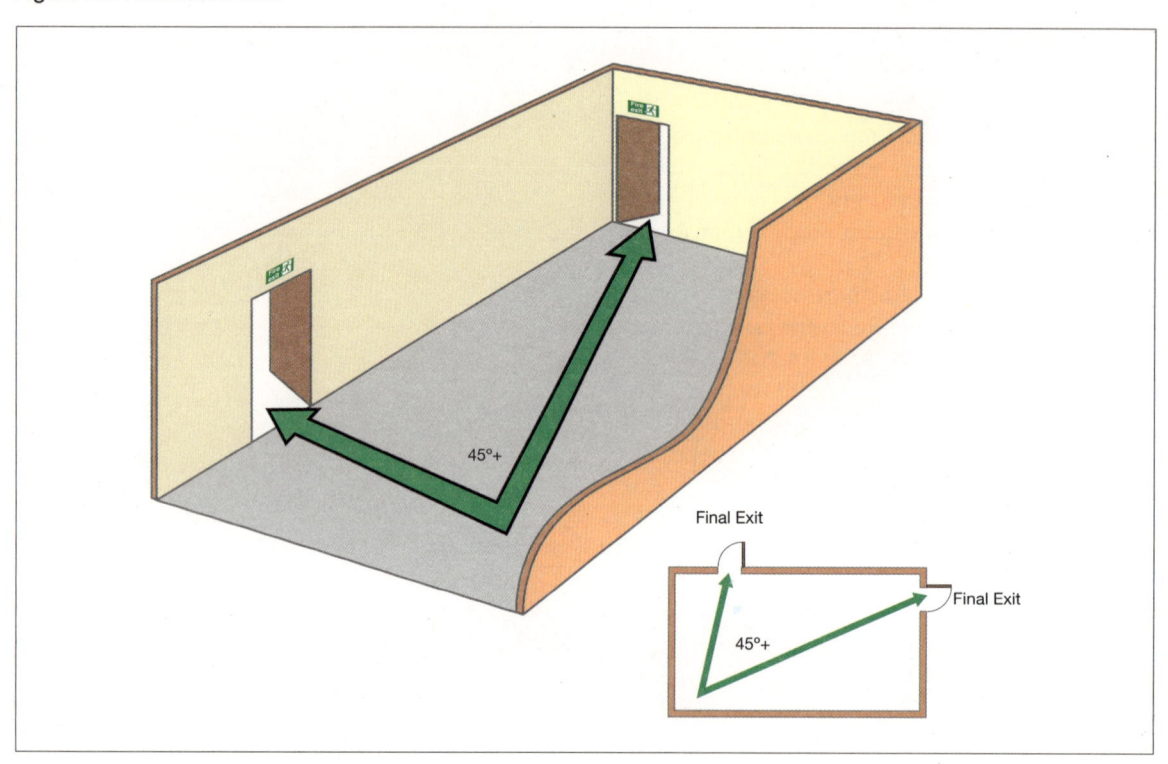

Figure 24: Alternative exits (separated by fire-resisting construction)

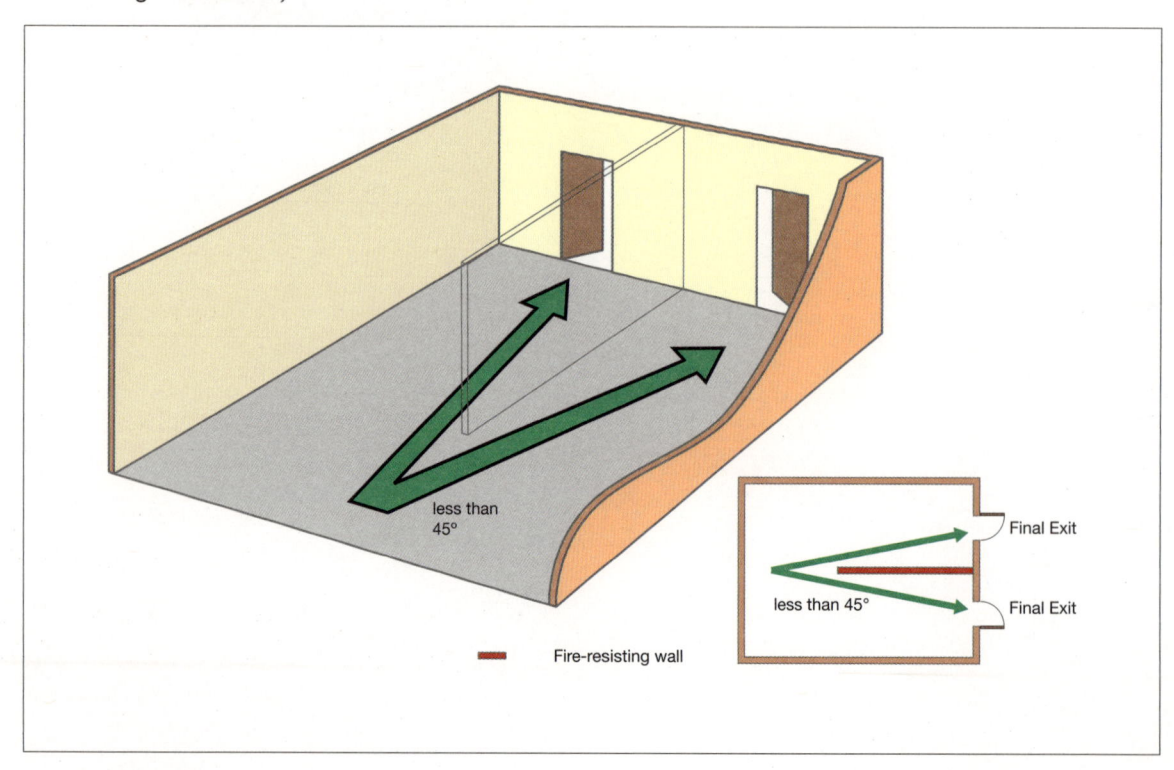

Figure 25: Inner rooms

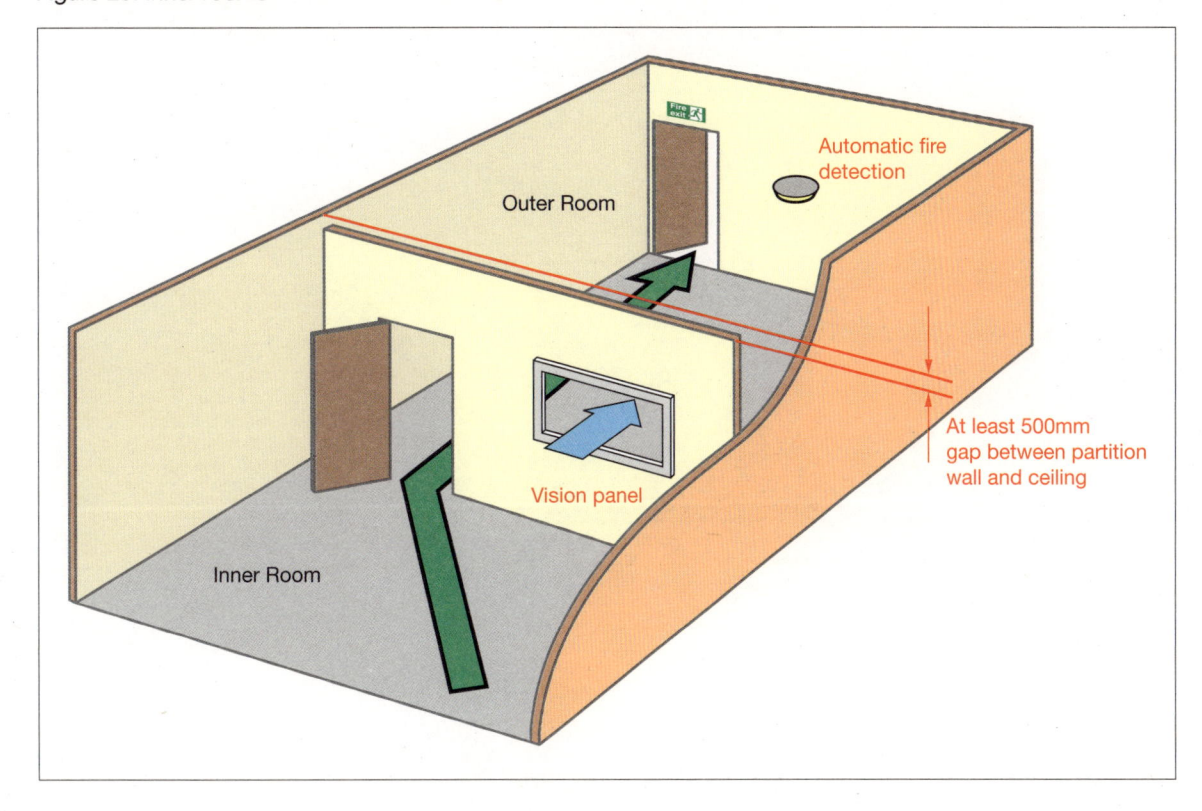

Inner rooms

Where the only way out of a room is through another room, an unnoticed fire in the outer room could trap people in the inner room. This means of exit should be avoided where possible. If, however, this cannot be achieved then adequate warning of a fire should be provided by **any one** of the following means:

- a vision panel between the two rooms providing adequate vision to give an indication of the conditions in the outer room and the means of escape;

- a large enough gap between the dividing wall and the ceiling, e.g. 500mm, so that smoke will be seen; or

- an automatic smoke detector in the outer room that will sound a warning in the inner room.

In addition, the following points should also be considered:

- Restrict the number of people using an inner room to 60.

- Access rooms should be under the control of the same person as the inner room.

- The travel distance from any point in the inner room to the exit from the access room should be restricted to escape in one direction only (see Table 2 on page 76), unless there are alternative exits from the access room.

- No one should have to pass through more than one access room while making their escape.

- The outer room should not be an area of high fire risk.

- The inner room should only be used as sleeping accommodation if smoke detection is provided in the access room.

Figure 26: Measuring travel distance from initial dead end (open plan)

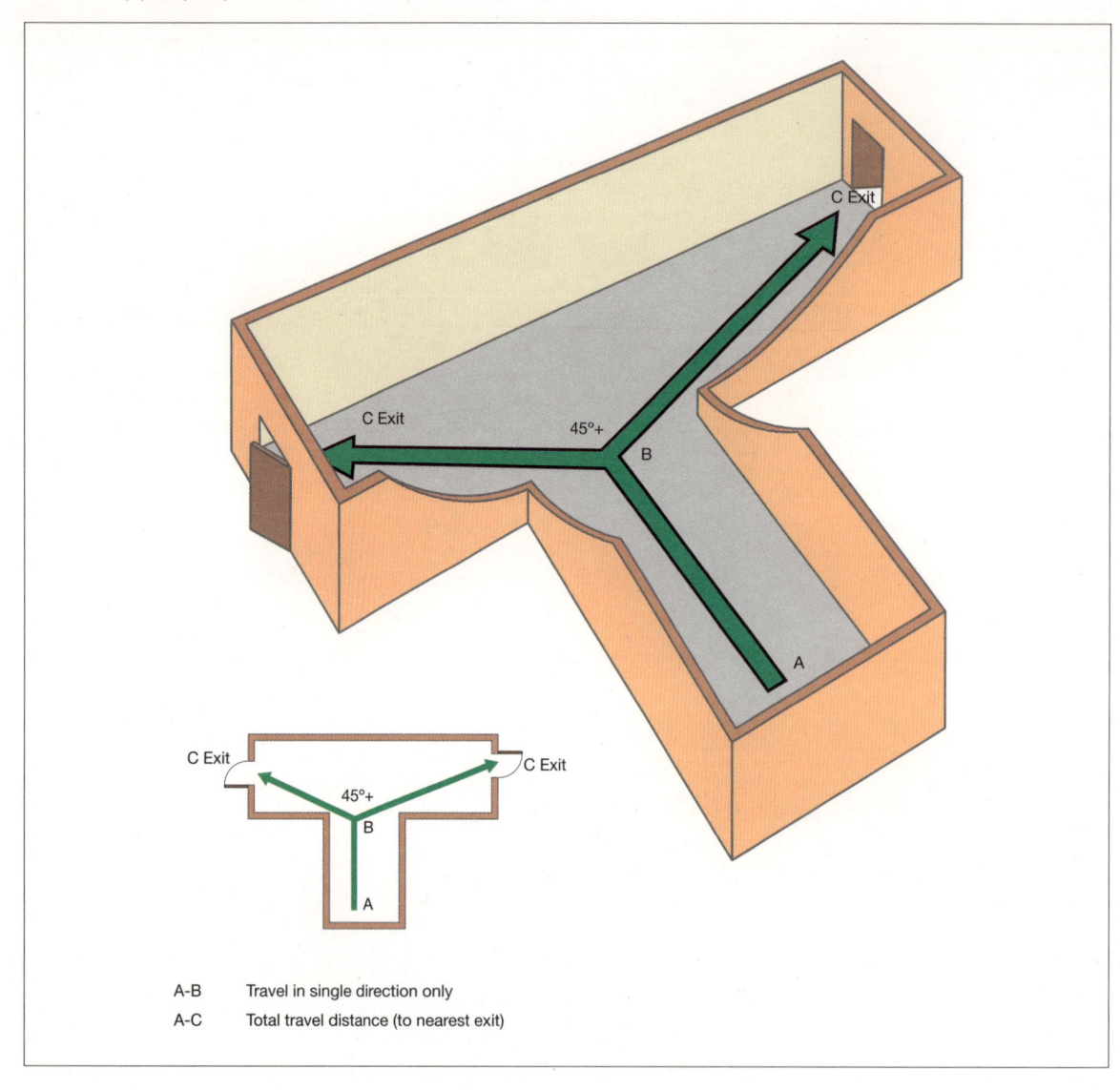

A-B   Travel in single direction only
A-C   Total travel distance (to nearest exit)

Measuring travel distances for initial dead end travel

Where the initial direction of travel in an open area (see Figure 26) or within an inner room (see Figure 27) is in one direction only the travel distance should be limited to that for a 'single escape route' in Table 2 on page 76. Any alternative exits should be positioned to ensure a fire will not compromise both exits. The maximum total travel distance recommended in Table 2 should apply to the nearest exit (Figures 26 and 27, distance A–C), your total travel distance should not exceed that for 'more than one escape route' in Table 2 on page 76.

Figure 27: Measuring travel distance from initial dead end (inner room)

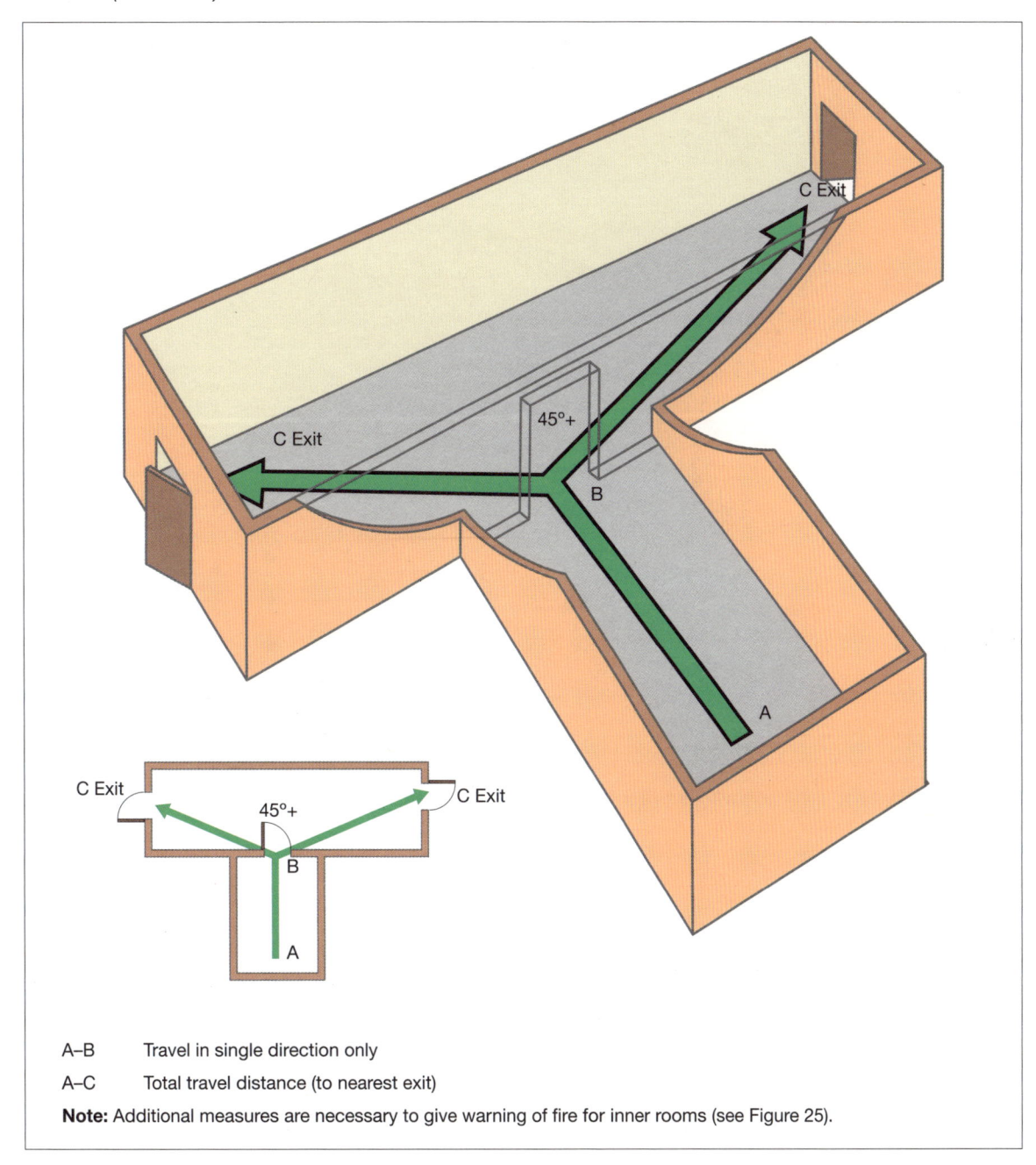

A–B	Travel in single direction only
A–C	Total travel distance (to nearest exit)

Note: Additional measures are necessary to give warning of fire for inner rooms (see Figure 25).

Escape routes with dead end conditions

If your premises has escape routes from which escape can be made in one direction only (a dead end), then an undetected fire in that area could affect people trying to escape. To overcome this problem, limit the travel distance (see Table 2 on page 76) and use one of the following solutions.

In non-sleeping areas (separate buildings or self-contained parts of buildings only):

- Fit an automatic fire detection and warning system in those areas where a fire could present a risk to the escape route, if not already in place (see Figure 28).

- Construct the exit route of fire-resisting partitions and self-closing fire doors to allow people to escape safely past a room in which there is a fire (see Figure 29).

- Provide an alternative exit (see Figure 30).

Alternative approaches may be acceptable, (e.g. an automatic fire suppression system), in which case you should seek advice from a competent person.

Figure 28: Dead end condition with automatic fire detection (non sleeping areas)

A–B	Travel in single direction only
A–C	Total travel distance
⬭	Automatic fire detection

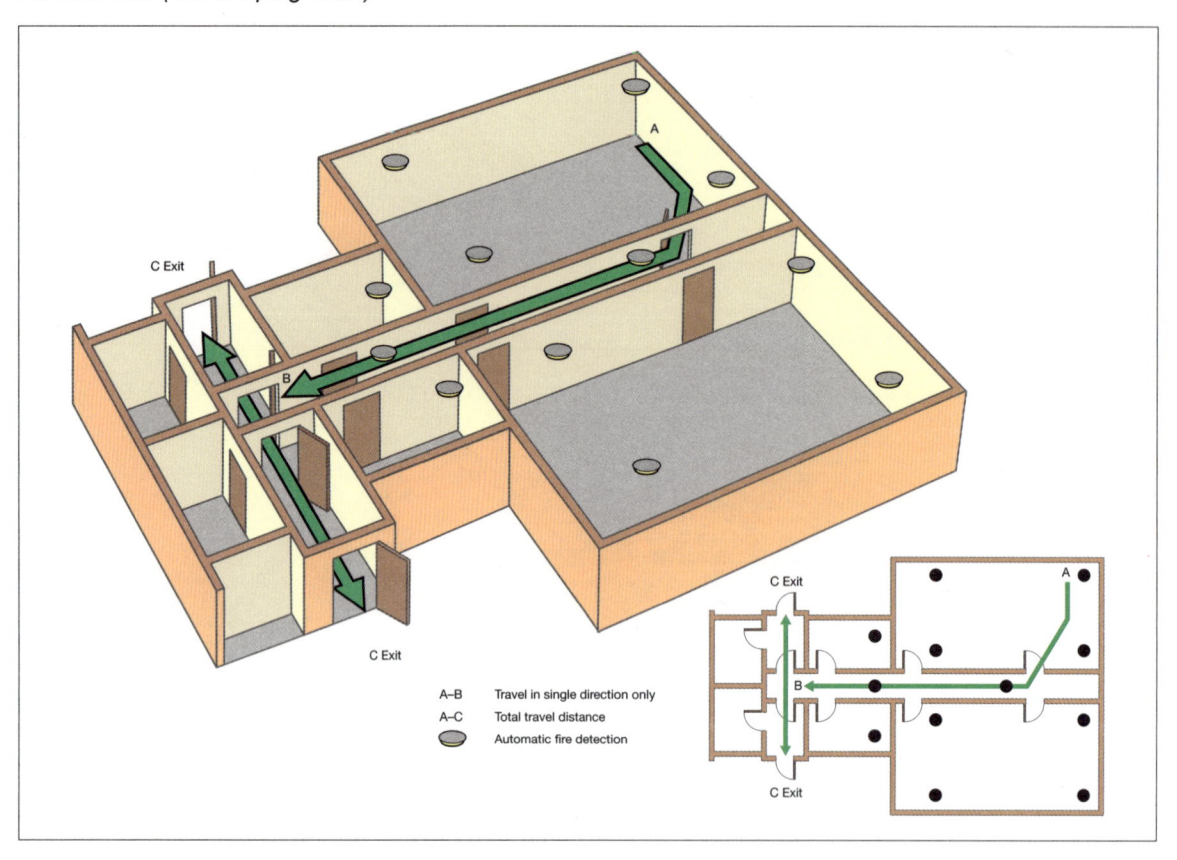

Figure 29: Dead end condition with fire-resisting construction (non sleeping areas)

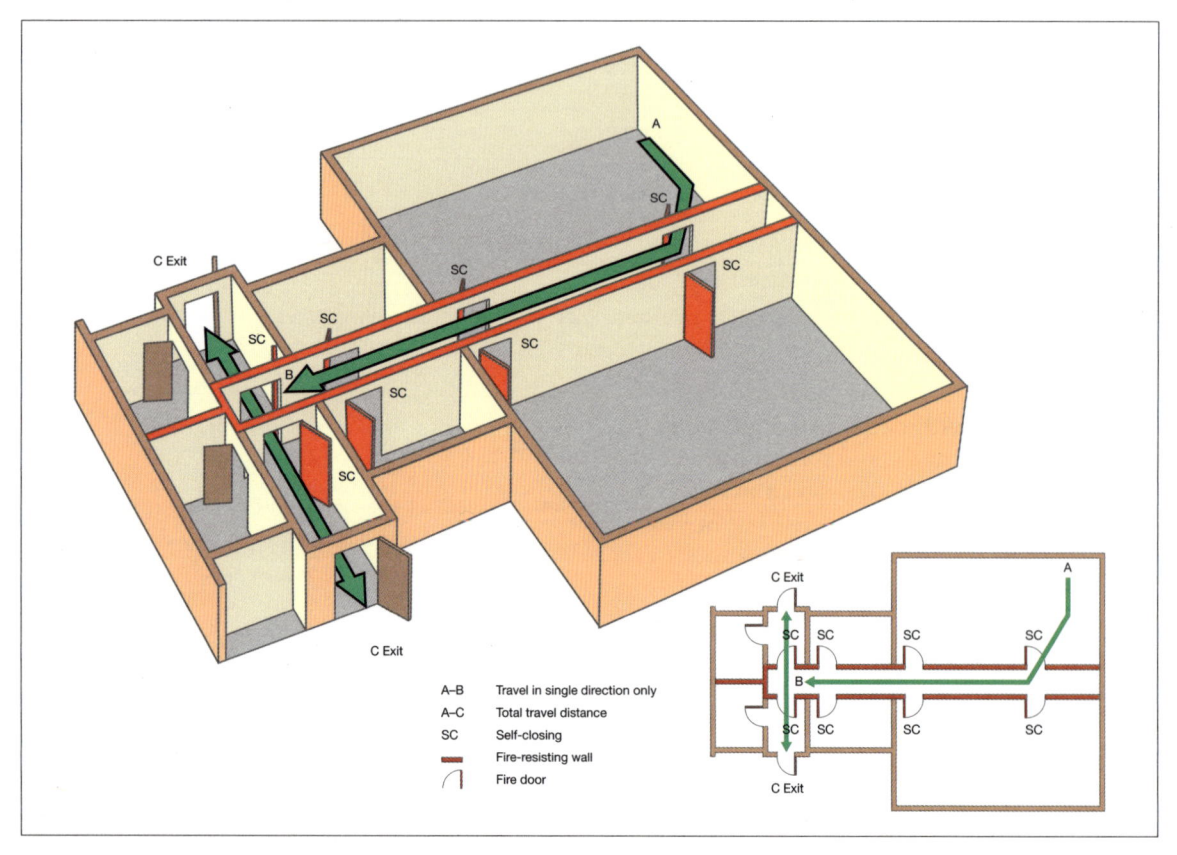

A–B	Travel in single direction only
A–C	Total travel distance
SC	Self-closing
▬	Fire-resisting wall
⌒	Fire door

Figure 30: Dead end condition provided with an alternative exit (non sleeping areas)

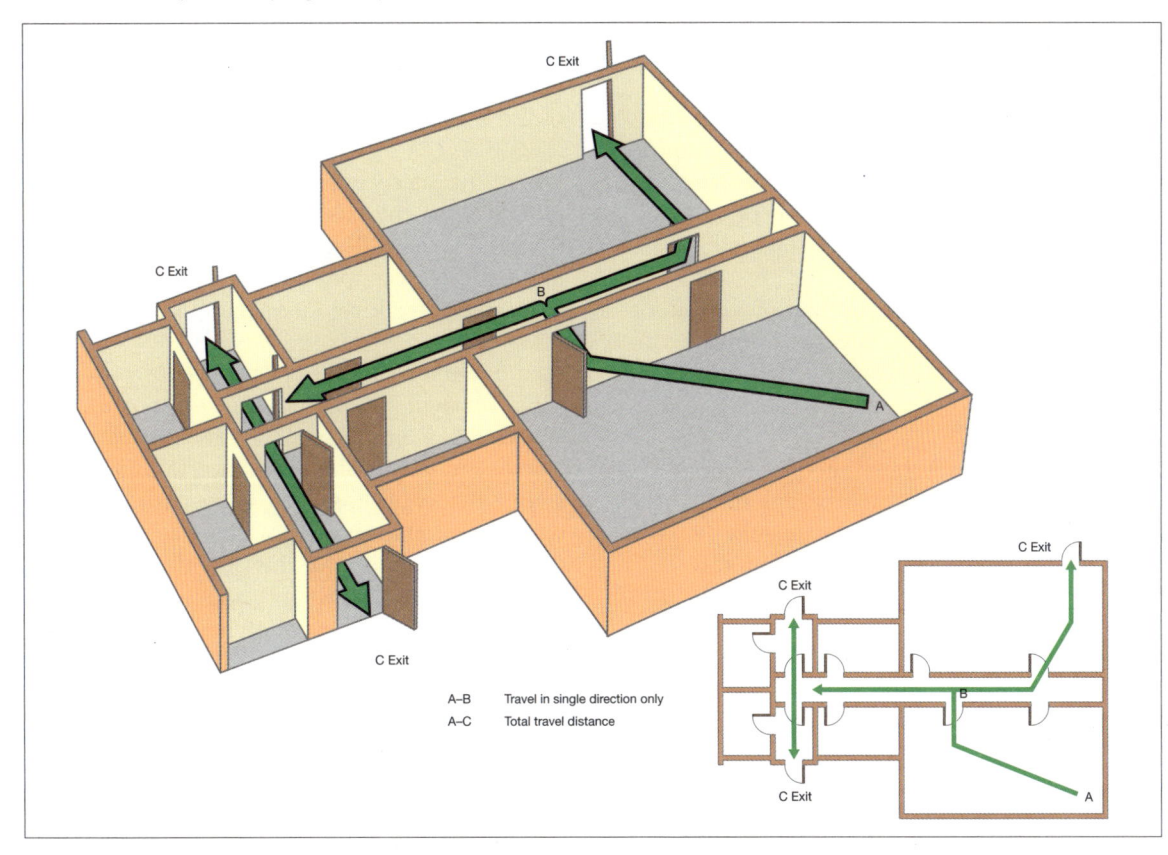

A–B Travel in single direction only
A–C Total travel distance

In sleeping areas the layout shown in Figure 31 will generally be acceptable. However, an alternative approach may be acceptable if other fire protection measures are in place in which case you should seek advice from a competent person.

Figure 31: Dead end with fire resising construction and automatic fire detection (in sleeping areas)

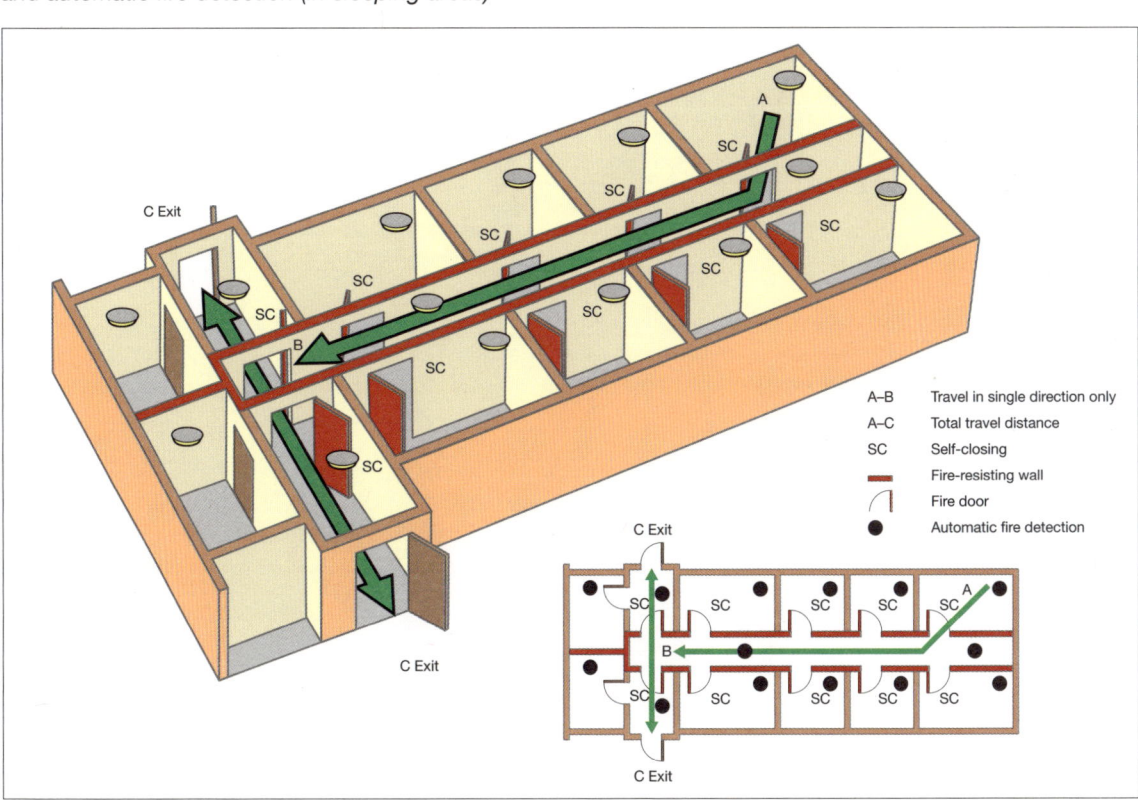

A–B Travel in single direction only
A–C Total travel distance
SC   Self-closing
▬   Fire-resisting wall
⌒   Fire door
●   Automatic fire detection

Part 2 • Further guidance on fire risk assessment and fire precautions

83

Basements, escape and protection

In all buildings with basements (other than small premises), stairways serving upper floors should preferably not extend to the basement and in any case should not do so where they are the only stairway serving the upper floors. Any stairway that extends from the basement to the upper floors should be separated at basement level by a fire-resisting lobby or corridor between the basement and the stairway.

All basements used by more than 60 people or where there are no exits directly to a place of total safety, should have at least two protected escape stairways.

In high risk premises these should be an alternative stairway from the basement to ground level, unless there is a suitable alternative route to the final exit from the basement.

Wherever possible all stairways to basements should be entered at ground level from the open air, and should be positioned so that smoke from any fire in the basement would not obstruct any exit serving the other floors of the building.

Where the stairway links a basement with the ground floor, the basement should be separated from the ground floor, preferably by two 30 minute fire-doors, one at basement and one at ground floor level (see Figure 32).

As a minimum, any floor over a basement should provide 60 minutes fire resistance. Where this is impractical, and as long as no smoke can get through the floor, automatic smoke detection linked to a fire-alarm system which is audible throughout the premises could, as an alternative, be provided in the basement area. If in doubt, contact a competent person for more detailed advice.

Figure 32: Basement protection

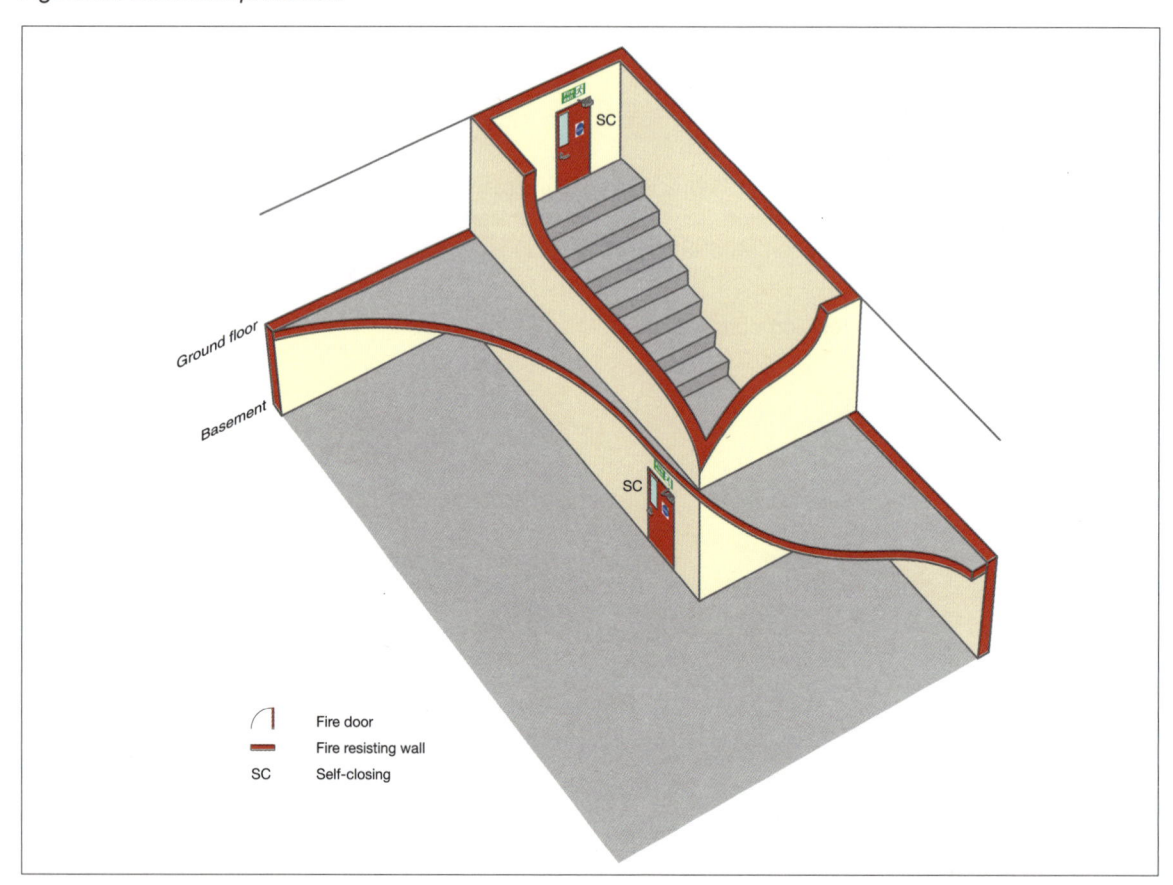

Protection of corridors

All corridors that serve sleeping accommodation should be of 30 minutes fire-resisting standard and fitted with self-closing devices.

Self-closing doors (particularly in corridors) may be an obstacle to the residents during the daily, normal, use of your home. To avoid the risk of doors being wedged open you may wish to install controlled door closing devices with an electromagnetic 'hold open' function which allow the door to close when the fire alarm actuates (see Appendix B2). Where fitted such doors should be closed at night.

Doors to cupboards, service ducts and any vertical shafts linking floors, and marked 'Fire door – keep locked shut when not in use'.

If your premises has corridors more than 30m long, then generally these corridors should be subdivided with fire doors as close to the mid point as possible, and, where necessary, fire-resisting construction to limit the spread of fire and smoke and to protect escape routes if there is a fire.

Corridors being used for horizontal evacuation should be subdivided with fire doors and fire-resisting construction to form protected areas (see Figure 20).

Where a corridor serves two exits from a floor, generally these corridors should be subdivided with fire doors to separate the two exits (see Figure 33).

Doors that are provided solely for the purpose of restricting the travel of smoke need not be fire doors, but will be suitable as long as they are of substantial construction, are capable of resisting the passage of smoke, and are self-closing. However, doors that subdivide a corridor to form a protected area should be fire-resisting. Smoke should not be able to bypass corridor doors, e.g. above a false ceiling, or via alternative doors from a room, or adjoining rooms, opening on either side of the subdivision.

Generally, false ceilings should be provided with barriers or smoke stopping over any fire doors. Where the false ceiling forms part of the fire-resisting construction this may not be necessary.

Depending on the outcome of your risk assessment, as an alternative to providing physical fire protection to the corridor, there may be other ways in which you could achieve an equivalent level of protection such as the combination of a suitable automatic fire suppression system used in conjunction with well designed sub-compartments (fire protected areas). You should first seek advice from a competent person.

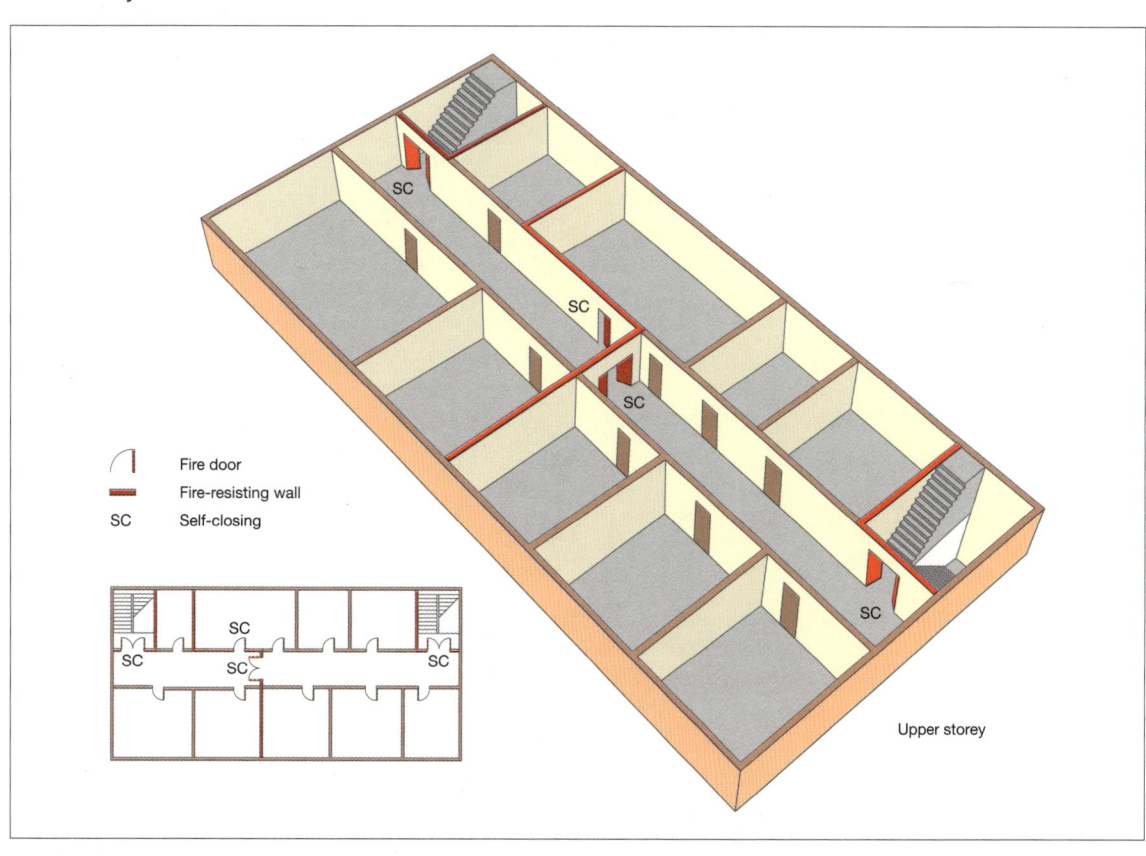

Figure 33: Subdivision of corridor between two stairways

Stairway enclosures

Stairways, if unprotected from fire, can rapidly become affected by heat and smoke, cutting off the escape route and allowing fire spread to other floors. However, if adequately protected, escape stairways can be regarded as places of reasonable safety to enable people to escape to a place of total safety.

In most premises designed and built to Building Regulations and served by more than one stairway, it is probable that these stairways will be protected by fire-resisting construction and will lead to a final exit. If any floor has an occupancy of over 60 each storey should have at least two exits i.e. protected routes. The figure of 60 can be varied in proportion to the risk, with lower risk the figure may be slightly increased, and with higher risk lower numbers of persons may be appropriate.

The benefit of protecting stairways from the effects of fire is that it allows you to measure your travel distance from the farthest point on the relevant floor to the nearest storey exit rather than the final exit of the building.

If you have a protected stairway(s) then it is essential that you maintain that level of fire protection.

It is possible that you may have some stairways which have no fire protection to them. In this case they are not designed for escape and are normally known as accommodation stairways (see accommodation stairways on page 90).

If you do not have a protected stairway and, depending on the outcome of your fire risk assessment, it may be that you can achieve an equivalent level of safety by other means. However, before doing so you should seek advice from a competent person.

If the building you occupy has floors which are occupied by different organisations to your own you need to consider, as part of your fire risk assessment, the possibility that a fire may occur in another part of the building over which you may have no control and which may affect the protected stairway if allowed to develop unchecked. If your fire risk assessment shows that this may be the case and people using any floor would be unaware of a developing fire, then additional fire protection measures may be required, e.g. an enhanced automatic fire detection and warning system. If this is required you will need to consult and co-operate with other occupiers and building managers.

You may find that stairways in your building are provided with protected lobbies or corridors at each floor level, except the top floor (see Figure 34). Although these are not generally necessary for means of escape in multi-stairway buildings of less than 18m high, they may have been provided for other reasons (e.g. firefighting access). In all cases protected corridors, lobbies and stairways must be kept clear of combustibles and obstructions.

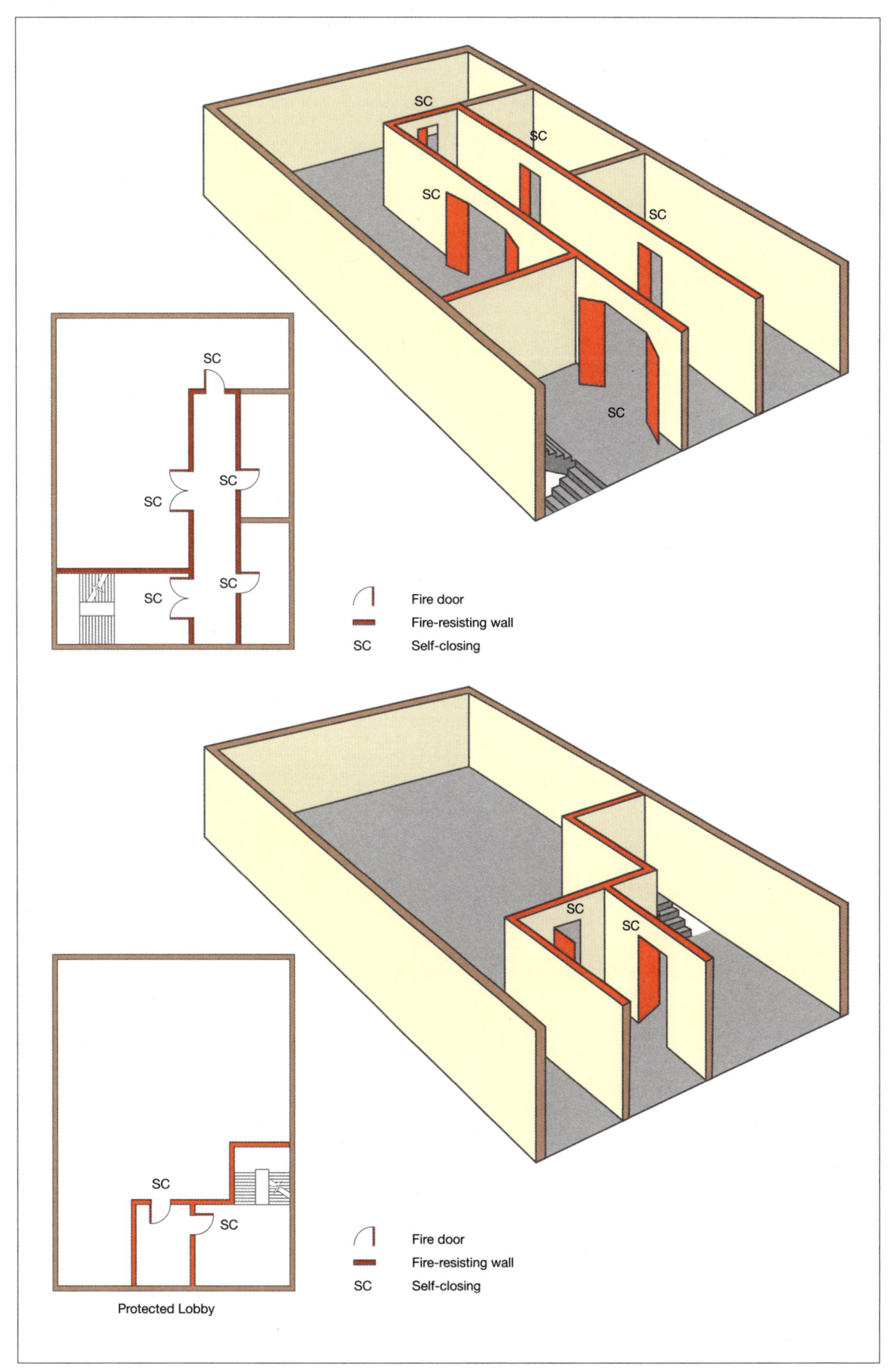

Figure 34: Examples of a stairway with protected lobby/corridor approach

Fire door
Fire-resisting wall
SC Self-closing

Protected Lobby

Fire door
Fire-resisting wall
SC Self-closing

Figure 34: Examples of a stairway with protected lobby/corridor approach

Ideally stairway enclosures should lead directly to a final exit. If your premises has only one stairway from the upper floor(s) which does not lead directly to a final exit, adopt one of the following arrangements:

- provide a protected route from the foot of the stairway enclosure leading to a final exit (see Figure 35); or

- provide two exits from the stairway, each giving access to a final exit via routes which are separated from each other by fire-resisting construction (see Figure 36).

Figure 35: Examples of a protected route from a stairway to a final exit

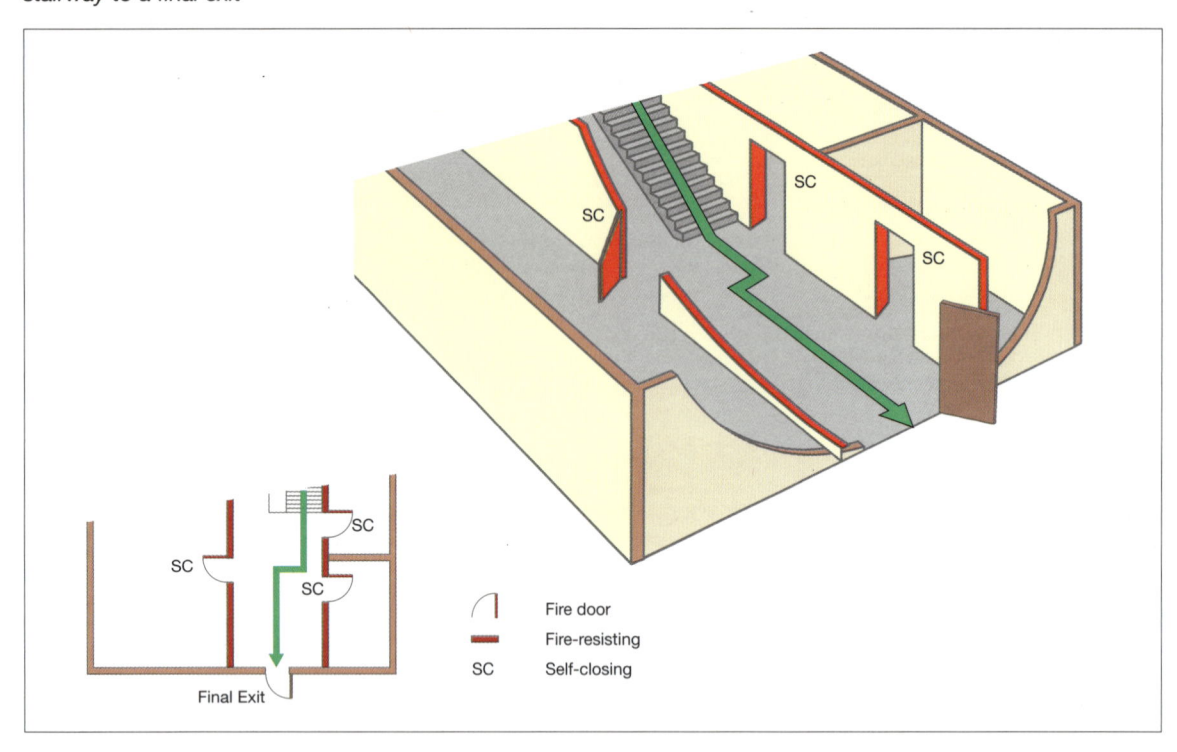

Figure 36: Examples of two escape routes from a stairway to final exits

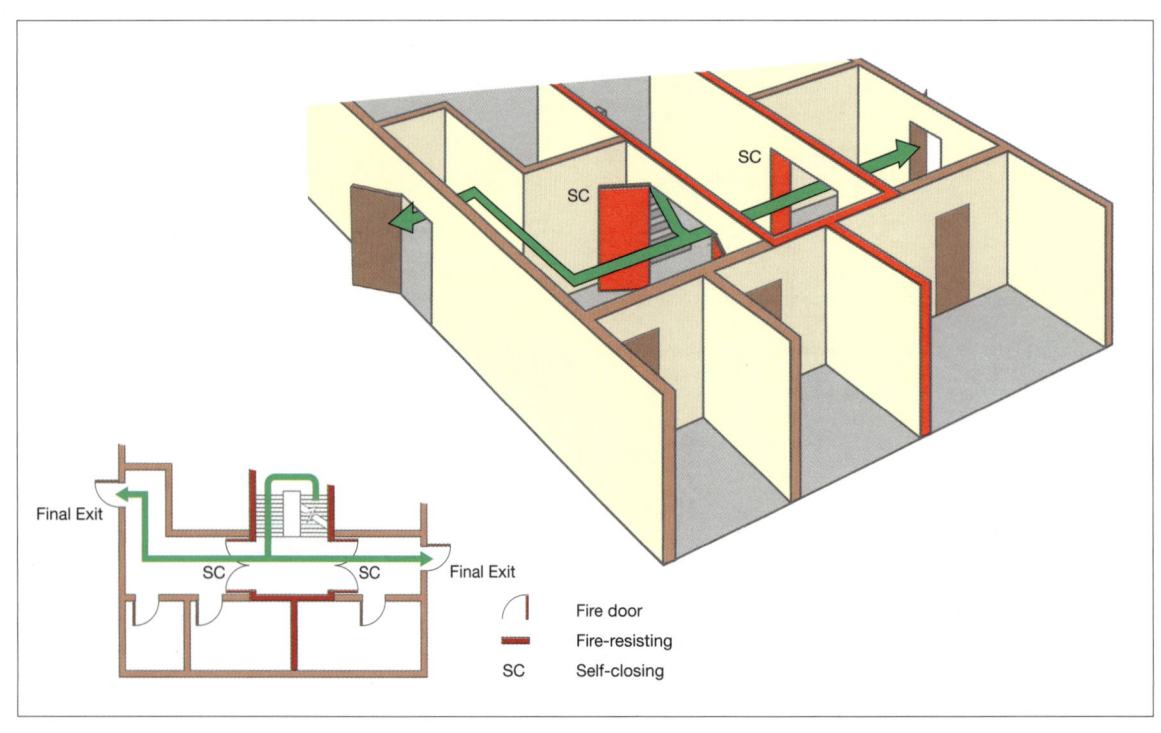

Separation of protected stairways

Where there are two or more protected stairways, the routes to final exits should be separated by fire-resisting construction so that fire cannot affect more than one escape route at the same time as illustrated in Figure 37.

Figure 37: Separation of protected stairway

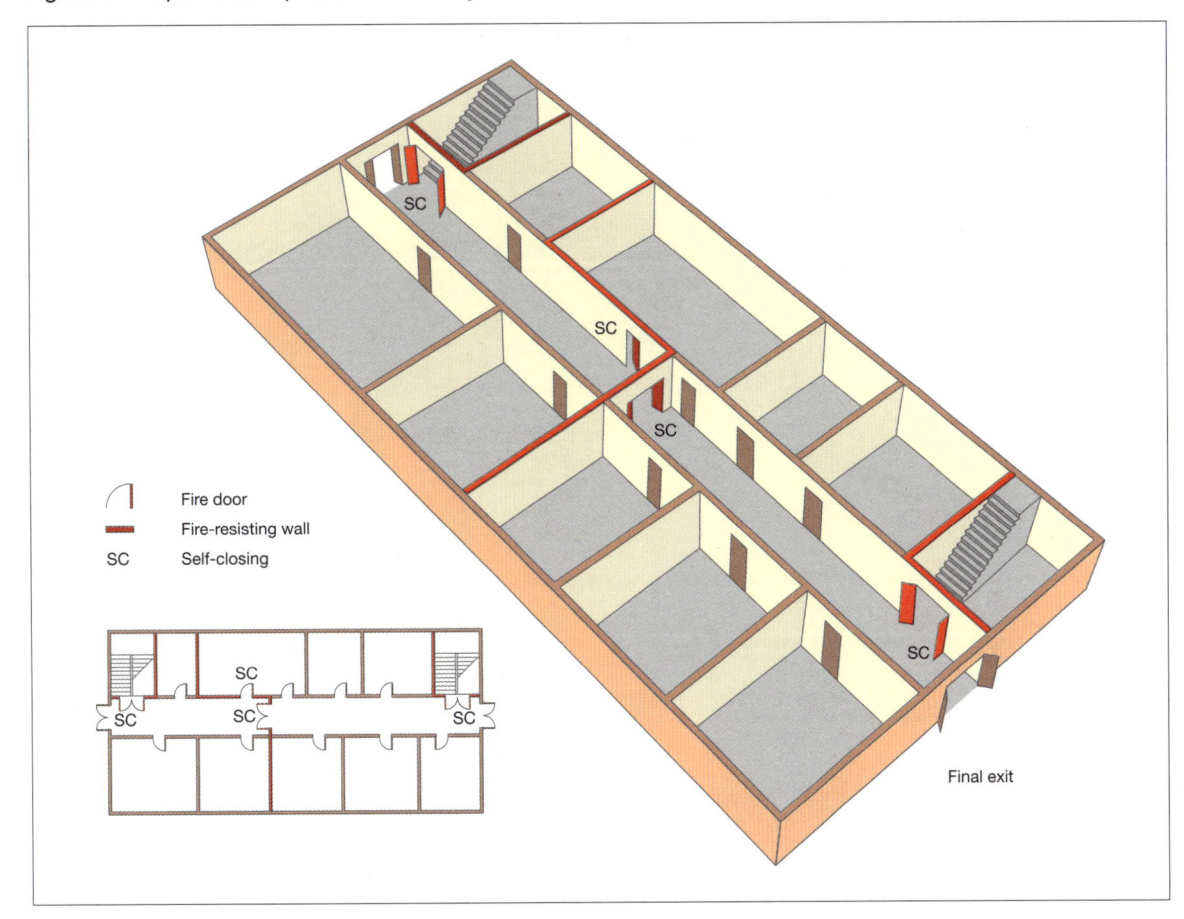

Creating a stairway bypass route

No one should have to pass through a protected stairway to reach another stairway. Options to avoid this include:

- using intercommunicating doors between rooms adjacent to the stairway, such doors must be available at all times when the building is occupied (see Figure 38);
- using balconies and other features to bypass the stairway;
- as long as there is enough space, create a bypass corridor around the stairway enclosure.

If you will be moving residents in wheelchairs you will need to ensure that there is sufficient width throughout the bypass route.

Note: Bypass routes should not be provided through bedrooms.

Figure 38: A stairway bypass route

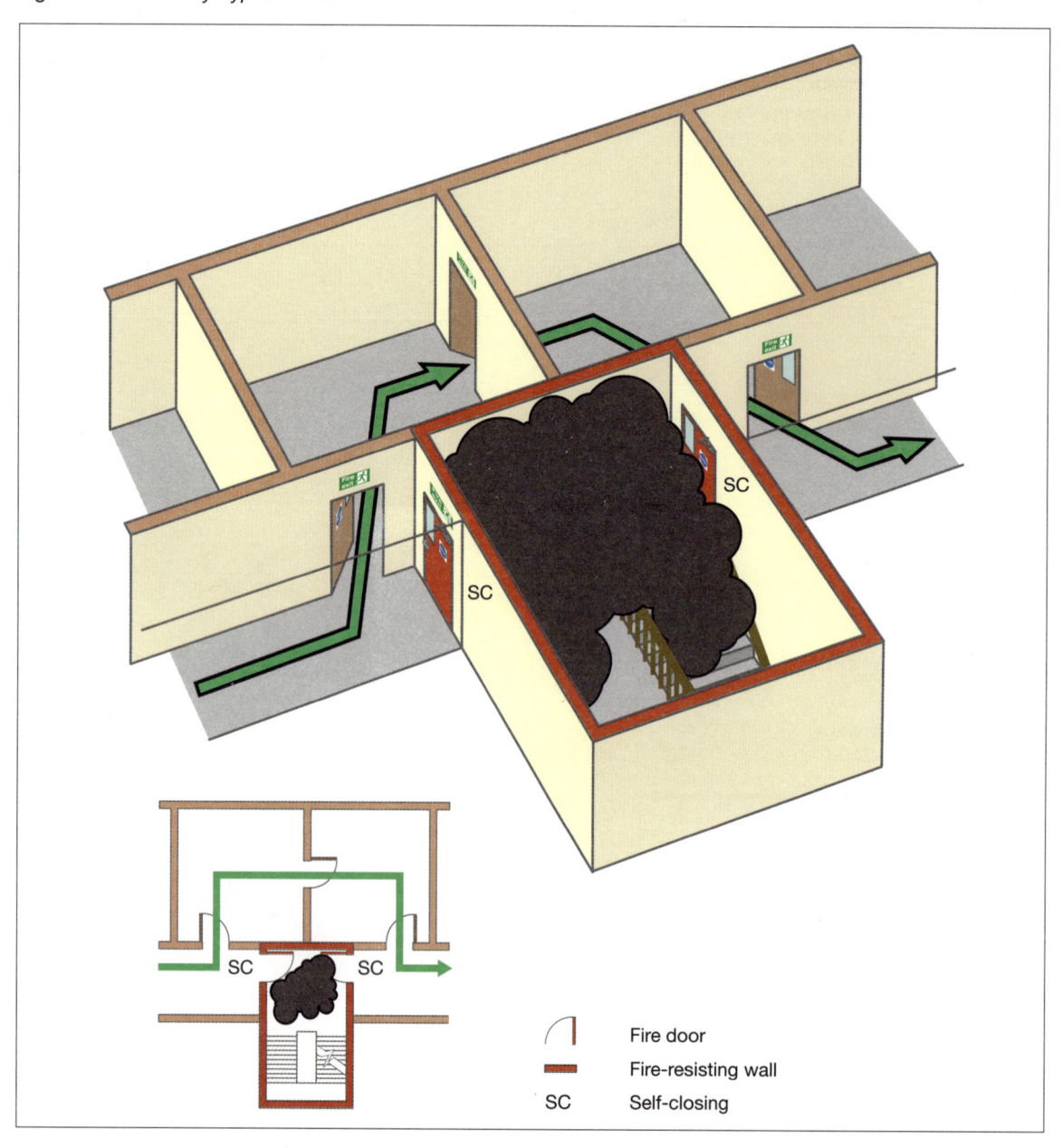

Reception areas

Reception or enquiry areas should only be located in protected stairways where the stairway is not the only protected one serving the upper floors, the reception area is small (less than 10m²) and is of low fire risk.

Accommodation stairways

If you have stairways that are used for general communication and movement of people in the premises, and they are not designated as fire escape stairs then these are called 'accommodation stairways'. They may not require fire separation from the remainder of the floor, as long as they do not pass through a fire compartment floor, or people do not have to pass the head of such a stairway in order to access a means of escape stairway. However, experience shows that many people will continue to use these as an escape route.

Accommodation stairways (or chair lifts, escalators and travelators) should not normally form an integral part of the calculated escape routes, however, where your fire risk assessment indicates that it is safe to do so, then you may consider them for that purpose. In these circumstances it may be necessary to seek advice from a competent person to verify this.

External stairways

To be considered a viable escape route, an external stairway should normally be protected from the effects of a fire along its full length. This means that any door, window (other than toilet windows) and walls within 1.8m horizontally and 9m vertically below any part of the stairway should be fire-resisting. Windows should be fixed shut and doors self-closing (see Figure 39).

Consider protecting the external stairway from the weather as the treads may become slippery, e.g. due to algae, moss or ice. If this is not possible, you must ensure that the stairway is regularly maintained. Consider fixing non-slip material to the treads.

External stairways should not normally be used by residents or members of the public.

Figure 39: Protection to an external stairway

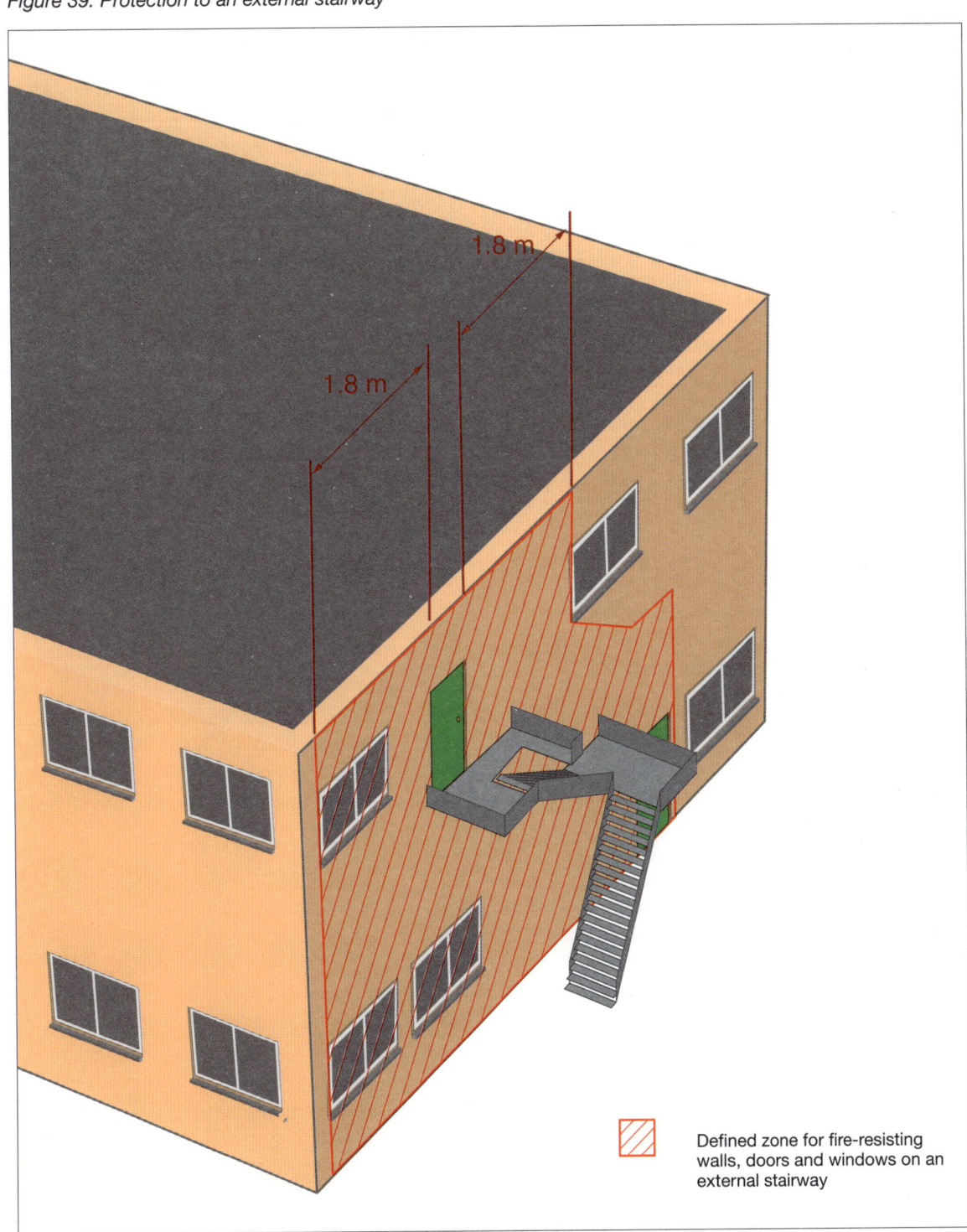

Spiral and helical stairways

Existing spiral and helical stairways that are not designed for the purpose should not normally form part of an escape route. They should not be used by residents, but, in exceptional situations they may be used by a small number of your staff. The stairway should not be more than 9m in height and not less than 1.5m in diameter with adequate headroom. A handrail should be continuous throughout the full length of the stairway.

However, spiral and helical stairways that have been designed for the purpose may be suitable for means of escape by able bodied staff, residents and the public. However, they are not usually suitable for young children. Further guidance is given in BS 5395-2.[26]

Roof exits

It may be reasonable for an escape route to cross a roof. Where this is the case, additional precautions will normally be necessary:

- The roof should be flat and the route across it should be adequately defined and well-illuminated where necessary with normal electric and emergency lighting. The route should be non-slip and guarded with a protective barrier.

- The escape route across the roof and its supporting structure should be constructed as a fire-resisting floor.

- Where there are no alternatives other than to use a roof exit, any doors, windows, roof lights and ducting within 3m of the escape route should be fire-resisting.

- The exit from the roof should be in, or lead to, a place of reasonable safety where people can quickly move to a place of total safety.

- Where an escape passes through or access another persons property you will need to ensure that you have a robust legal agreement in place to allow its use at all times when people are on your premises.

Roof exits should not normally be used by residents or members of the public.

External escape routes should receive routine inspection and maintenance to ensure they remain fit for use.

External escape routes should receive routine inspection and maintenance to ensure they remain fit for use.

If your premises are part of a multi-occupied building you will need to liaise with the owner or other persons responsible for the building. Where an escape routes passes through or across another persons property you will need to ensure that any legal agreements in place cover access for maintenance of the escape route.

A typical escape route across a roof is illustrated in Figure 40.

Figure 40: An escape route across a roof

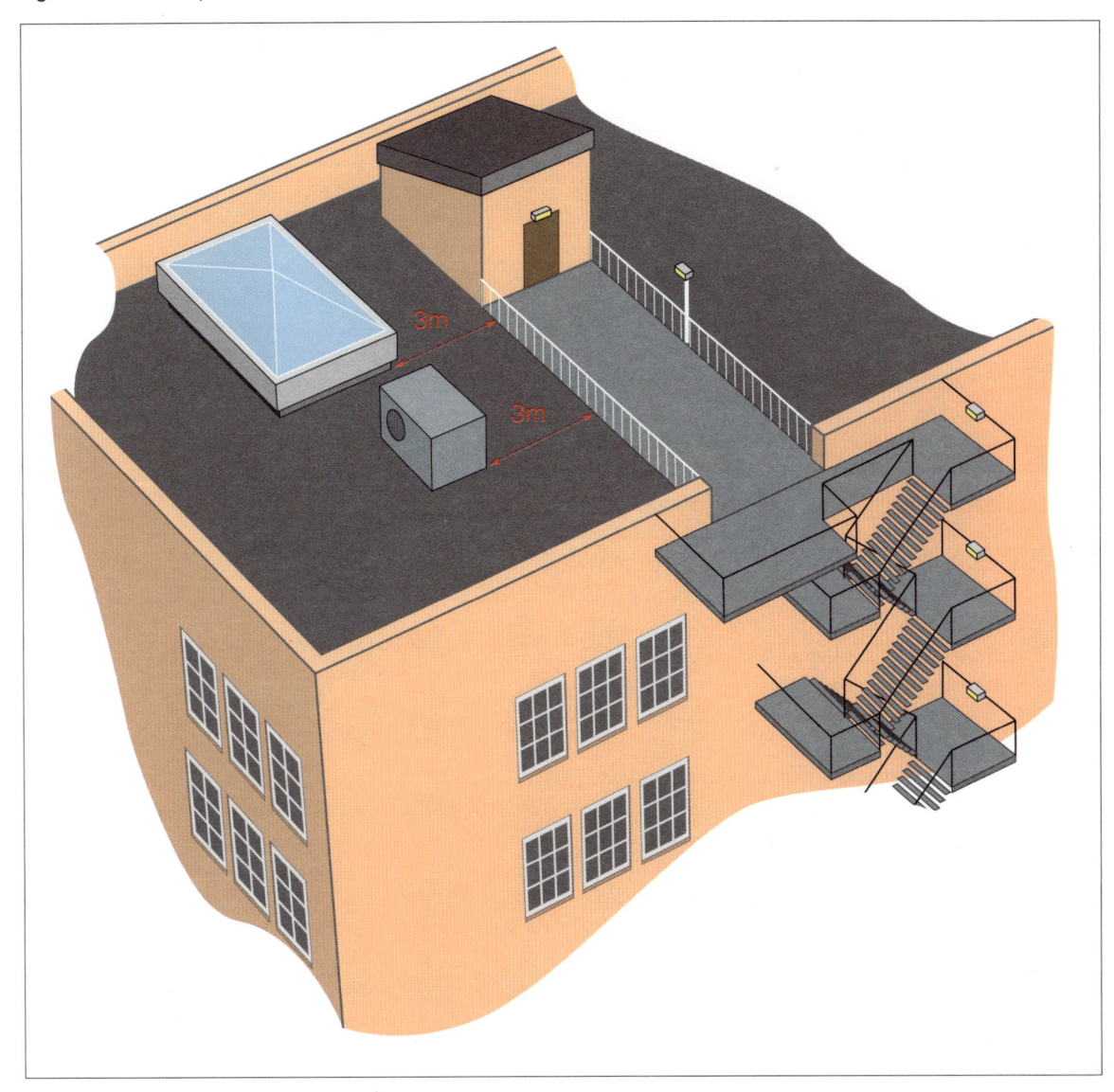

Revolving doors, wicket doors, roller shutters and sliding doors

Revolving doors, should not normally be considered as escape doors unless the leaves fold outward to form a clear opening upon pressure from within, or standard doors of the required exit width are provided next to the revolving door.

Ideally wicket doors or gates should have a minimum opening height of 1.5m. The bottom of the door should not be more than 250mm above the floor and the width should be preferably more than 500mm but not less than 450mm. Normally wicket doors will only be suitable for up to 15 members of staff however, in areas of a higher fire risk, this should be reduced to a maximum of three.

Loading and goods delivery doors, shutters (rolling, folding or sliding), up-and-over doors and similar openings are not normally suitable for use as a final exit. However, they may be suitable for escape from areas of normal risk by staff familiar with the escape routes and operation of the doors, as long as they are not likely to be obstructed and can be easily and immediately opened manually, (even if normally power-operated) and provided that the staff are familiar with that escape route.

Further information on controlled door closing devices, automatic door releases and fastenings is given in Appendix B.

Wall and floor hatches

Only in exceptional circumstances should it be necessary to rely for means of escape on wall hatches and floor hatches. However, there may be some instances when, because of structural difficulties, it will be reasonable to accept arrangements of this kind for a very limited number of persons who are active enough to use them, but under no circumstances should they be provided for the evacuation of residents. Where wall and floor hatches are provided, there may be a need to take special precautions to safeguard against their obstruction and protect their use, e.g. by the provision of guard-rails round the hatchway.

Lifts
Due to the danger of the power supplies to a lift being affected by a fire, lifts not specifically designed as 'firefighting' or 'evacuation' lifts are not normally considered acceptable as a means of escape. However, where a lift and stairway for a means of escape are incorporated in a fire-resisting shaft which has a final exit from it at the access level and the lift has a separate electrical supply to that of the remainder of the building, than that lift subject to an agreed fire risk assessment, may be acceptable as a means of escape in case of fire.

Lifts are housed in vertical shafts that interconnect floors and compartments, therefore precautions have to be taken to protect people from the risk of fire and smoke spreading from floor to floor via the lift shaft. Such precautions may include:

- Separating the lift from the remainder of the story using fire-resisting construction and access via a fire door;
- Ensuring the lift shaft is situated in a protected enclosure which may also be a stairway enclosure;
- Providing ventilation of at least $0.1m^2$ at the top of each lift-well so that any smoke can escape.

Final exit doors and escape away from the premises
Good escape routes to a final exit will be of little benefit if the occupants are not able to get out of the building and quickly disperse from the area to a place of total safety. It is also important to consider where people will go once they have evacuated from the premises.

The matters that you should consider include the following:

- Final exit doors should be quickly and easily openable without a key or code in the event of a fire. Where possible, there should be only one fastening. See Appendix B3 for more information on security fastenings.
- Final exit doors should not lead people into an enclosed area from which there is no further escape.
- Where a final exit discharges into an enclosed area, further access to a place of total safety should be available by means of further doors or gates that can be easily opened in a manner similar to the final exit.
- The route away from the premises needs to be clearly defined, illuminated, signposted, and if necessary protected by guardrails. Consider the proximity of other building outlets, such as ventilation and refuse shafts, which may emit smoke and heat from the fire.
- Dangers once away from the building e.g. busy roads and people traffic passing the building, that may affect people on their way out.

Your fire risk assessment must take into account factors such as the mobility of the people who may be expected to use the doors and the accessibility of door opening devices.

Portable ladders and throw-out ladders
Throw-out ladders are not acceptable as a means of escape. Portable ladders are only acceptable in exceptional circumstances where it will be reasonable for ladders of this kind to provide escape for one or two able-bodied staff (e.g. from a high level plant room).

Fixed vertical or raking ladders
These are normally only suitable for use of a limited number of able-bodied staff where they form the only means of access to plant spaces, provided they are suitably guarded and the total descent does not exceed 9m without an intermediate landing. Where they are external they should be protected from the elements in the same way as external stairways.

Lowering lines and other self-rescue devices
These are not acceptable as a means of escape.

Means of escape – security
It is accepted that in many care home situations conflicts between the needs for means of escape and security may arise, particularly in premises which provide accommodation for people with mental illness, where it may be essential to maintain a high level of supervision during an evacuation.

In areas where security is important a properly designed and managed interface between staffing levels, installed fire safety measures and security measures should be sufficient to permit staff controlled evacuation systems.

Any solution proposed, that includes security controls on escape routes, must be discussed and agreed with the relevant enforcing authorities, care provider and other relevant bodies.

4.2 Escape route layout

The examples listed in Table 3 show typical escape route solutions for a range of common building layouts. In each case the solution is for a normal risk building unless illustrated otherwise. These are not intended to be prescriptive or exhaustive but merely to help you understand how the principles of means of escape may be applied in practice.

They are illustrative of the key features of escape route layouts and not intended to be real building layouts or to scale.

You do not need to read all of this section, you only need to consider those diagrams which most closely resemble your premises. If your building does not resemble these then you should seek expert advice. These examples are intended to represent your existing layout; they are not to be used as design guidance.

In all of these examples the following basic principles apply:

- The farthest point on any floor to the final exit, storey exit or protected area is within the overall suggested travel distance (see Table 2 on page 76).
- The route to and the area near the exit is kept clear of combustibles and obstructions.
- The protected area or stairway is kept clear of combustibles and obstructions.
- The escape route leads to a final exit.
- Where the stairway is not protected, the final exit is visible and accessible from the discharge point of the stairway at ground floor level.
- High-risk rooms do not generally open directly into a protected area or stairway.
- If your fire risk assessment shows that there will be people using any floor who would be unaware of a fire you may require additional fire-protection measures, e.g. an extended automatic fire-detection and warning system.
- There should be more than one escape route from all parts of the premises (rooms or storeys) except for areas or storeys with an occupancy of less than 60. The figure of 60 can be varied in proportion to the risk, for a lower risk there can be a slight increase, for a higher risk, lower numbers of persons should be allowed.

Table 3: Typical examples of escape route layouts

Single-storey buildings with more than one exit	
Ground floor with more than one exit (cellular and open plan)	See Figure 41
Multi-storey buildings with more than one stairway	
Two-storey premises (non sleeping areas)	See Figure 42
Two-storey with protected corridor (sleeping areas)	See Figure 43
Three-storey (basement, ground and first floor)	See Figure 44
Three and four-storey (ground and up to three upper floors) horizontal evacuation	See Figure 45
Multi-storey building with a single stairway (in part of the building)	
Ground and up to three upper storeys – protected by lobbies/corridors (sleeping areas)	See Figure 46
Ground and up to three upper storeys – protected with automatic fire detection (non sleeping areas with very low risk)	See Figure 47

If you do not have any of the stairway configurations given, and depending on the outcome of your fire risk assessment it may be that you can achieve an equivalent level of safety by other means.

The green arrows on the Figures 41-47 represent the travel distances given in Table 2 on page 76 which should be applied.

If your building has ground and more than three upper storeys, ask advice from a competent person.

Single-storey buildings with more than one exit

Ground floor with more than one exit (cellular and open plan) Figure 41 shows examples of ground floor cellular and open plan layouts with more than one exit. In sleeping areas the corridor should be a protected route with fire-resisting construction and self-closing fire doors and protected by automatic fire detection. In non-sleeping areas the fire-resisting construction may be required to sub-divide the premises into fire protected areas.

Note: If any room or area in a non-sleeping area opens on to any part of the escape route from a sleeping area, automatic fire detection will be necessary in these areas.

Figure 41: Ground floor with more than one exit (cellular and open plan)

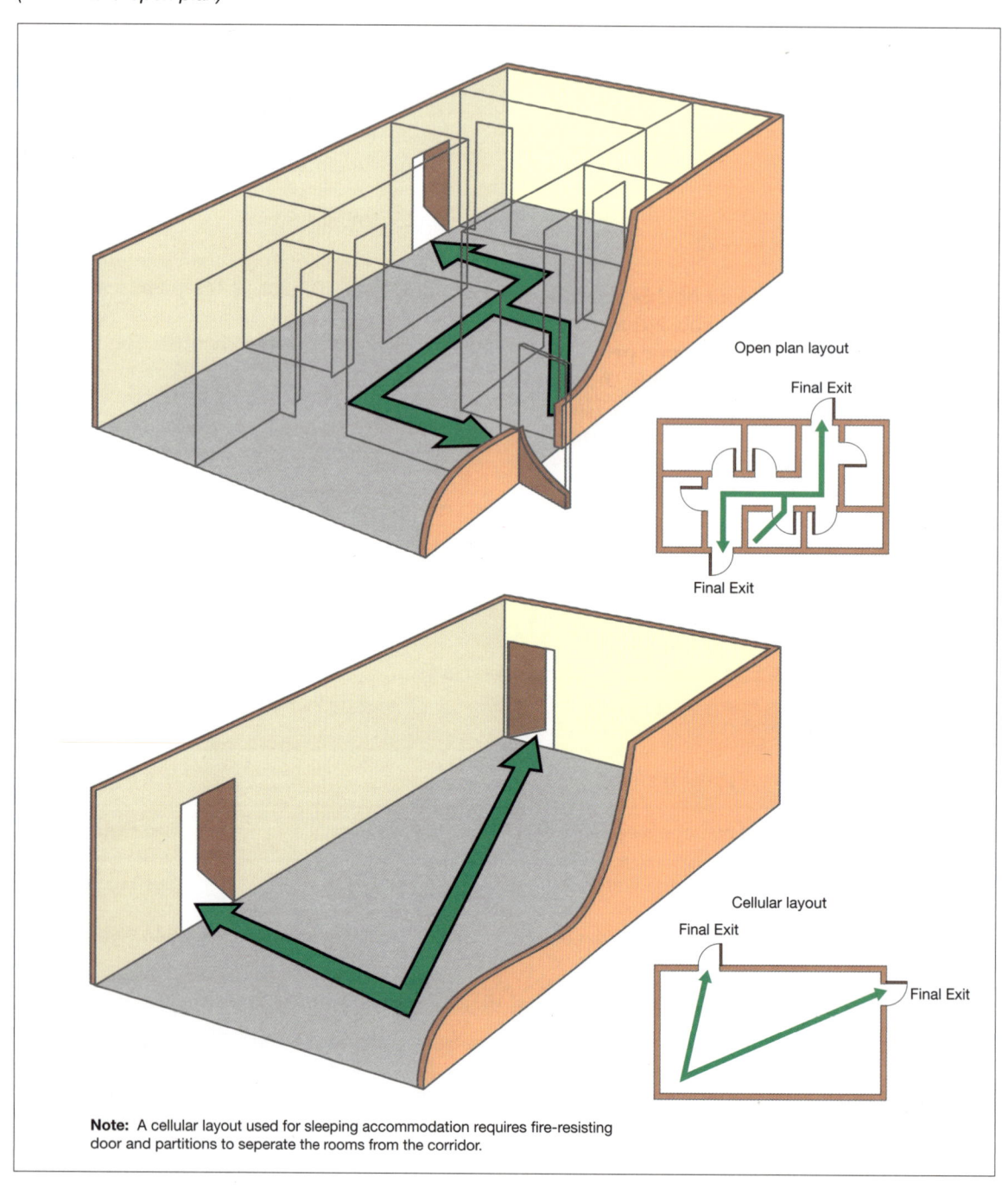

Note: A cellular layout used for sleeping accommodation requires fire-resisting door and partitions to seperate the rooms from the corridor.

Multi-storey building with more than one stairway

Two storey (non-sleeping and sleeping)

If your premises has a ground floor and one upper storey and these are served by more than one stairway, it is important to understand that you are unlikely to be able to meet the suggested travel distance to a final exit (see Table 2 on page 76). Therefore stairways may need to be protected by a fire-resisting enclosure as shown.

The layout shown in Figure 42 will be generally acceptable as long as the furthest point on each of your floors to the storey exit is within the overall suggested travel distance (see Table 2). This shows an unprotected corridor, in non sleeping areas.

Note: If any room or area in a non-sleeping area opens on to any part of the escape route from a sleeping area, automatic fire detection will be necessary in these areas.

The layout shown in Figure 43 shows a protected corridor, in sleeping areas.

Where necessary additional fire-resisting construction may be required to subdivide the premises into fire protected areas.

Figure 42: Two-storey (non sleeping areas)

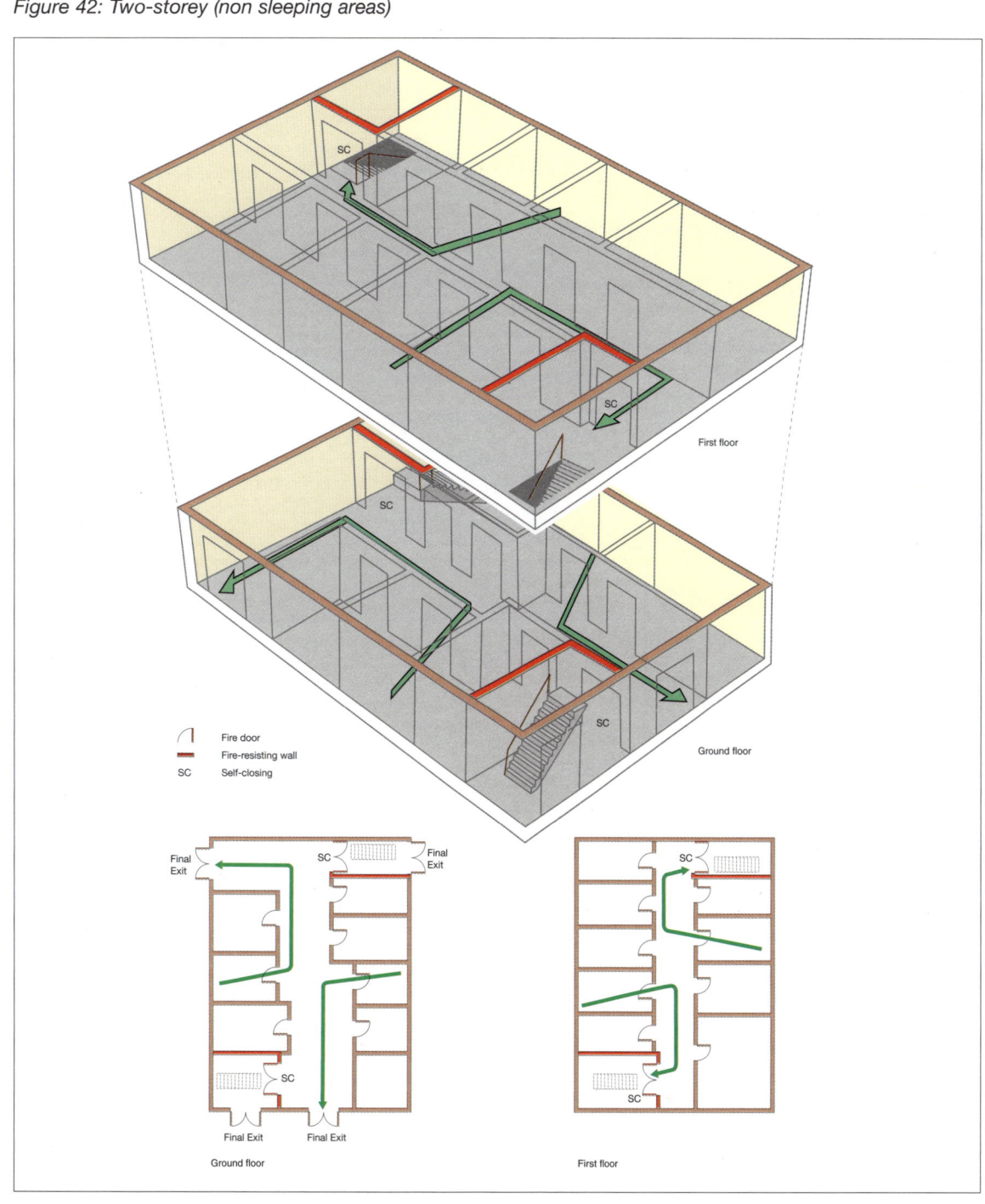

Figure 43: Two-storey with protected corridor (sleeping areas)

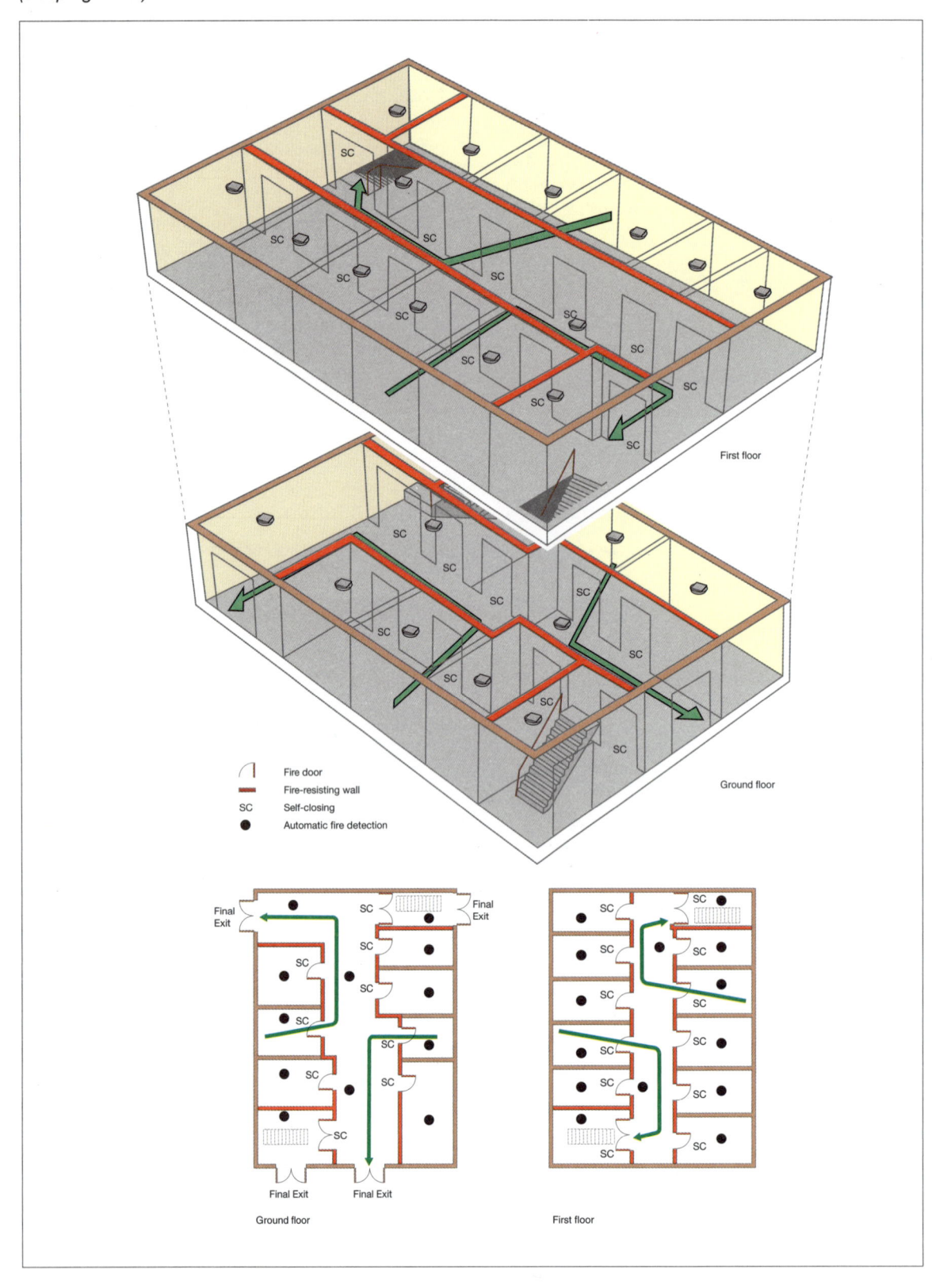

Three-storey (basement, ground and first floor)

In premises with a basement, ground and first floor, served by more than one stairway, the layout shown in Figure 44 will be generally acceptable as long as the following apply:

- To overcome the restriction of travel distance the stairway has been completely enclosed in 30-minute fire-resisting construction and all doors onto the stairway are self-closing fire doors.

- The furthest point on all of the floors to the nearest storey exit is within the overall suggested travel distance (see Table 2 on page 76).

- The stairways from the basement to ground floor level are separated by a fire-resisting lobby or corridor between that basement and the protected stairway.

- The floor between the ground floor and basement is fire-resisting, e.g. 60 minutes.

Note: Where necessary additional fire-resisting construction may be required to subdivide the premises into fire protected areas.

This principle applies to taller buildings (up to 18m). However, where your building has ground and more than three upper storeys ask advice from a competent person.

Figure shows distances to *nearest* stairway (or final exit).

Figure 44: Three-storey (basement, ground and first floor)

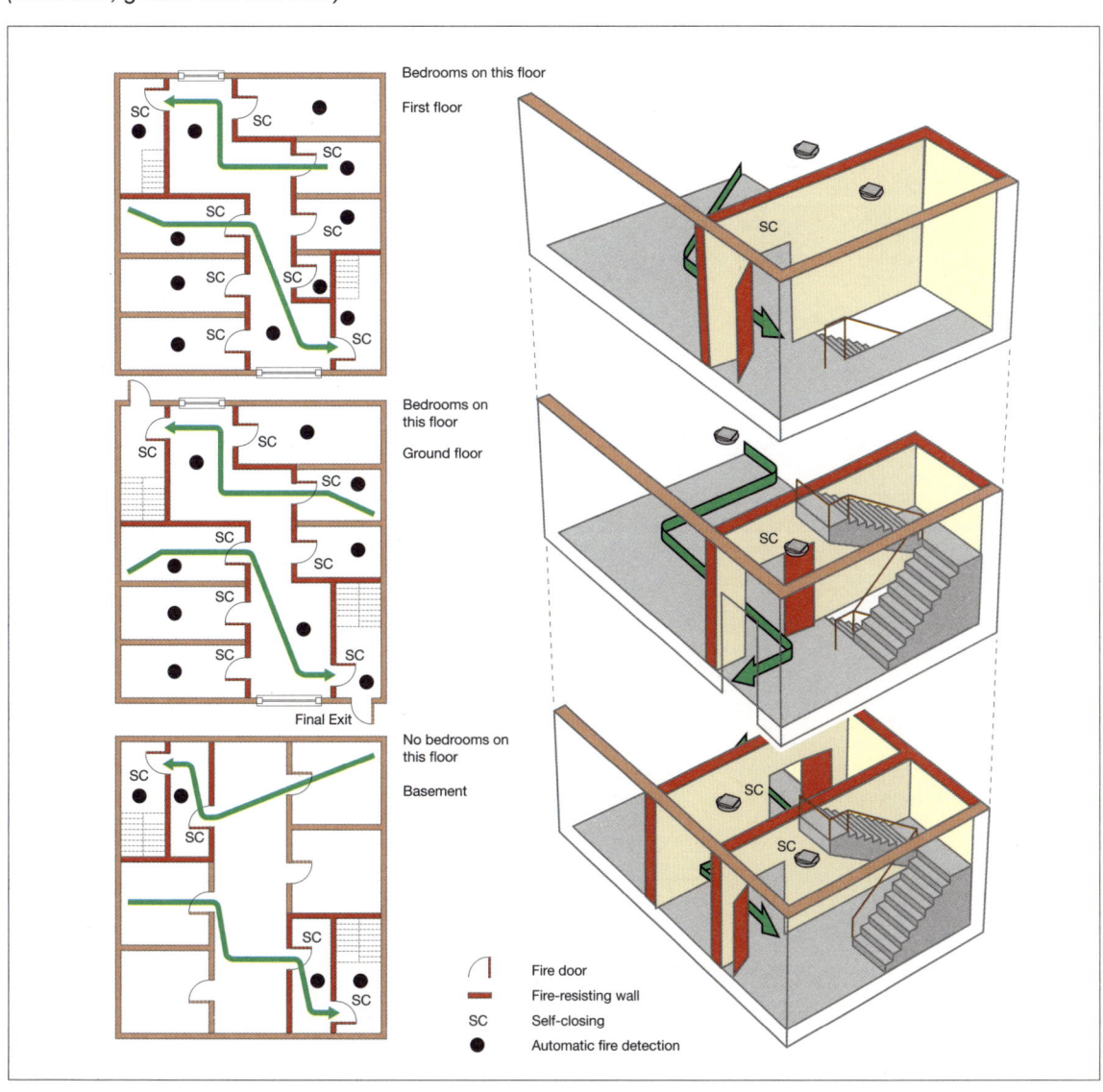

Three and four-storey (ground and up to three upper floors) horizontal evacuation

The layout shown in Figure 45 will be generally acceptable as long as the furthest point on all of your floors to the nearest protected area or to a storey exit is within the overall suggested travel distance (see Table 2 on page 76).

This principle applies to taller buildings (up to 18m). However, where your building has more than ground and three upper storeys ask advice from a competent person.

The Figure shows distances to *nearest* protected area or stairway (or final exit).

Figure 45: Three and four-storey (ground and up to three upper floors)

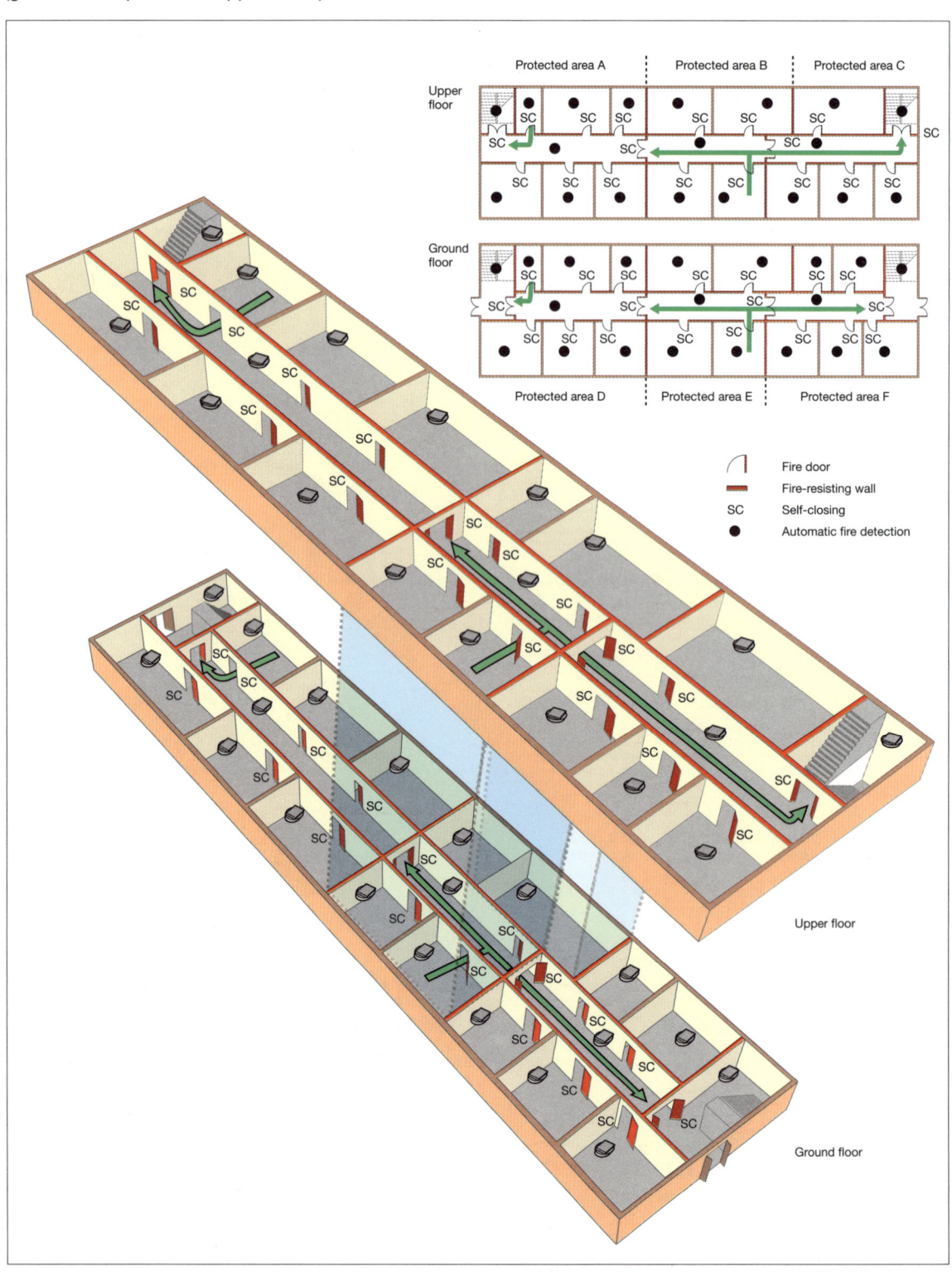

100

Multi-storey buildings with a single stairway (in part of the building)

In the majority of multi-storey premises, two or more stairways will be provided for escape purposes. Parts of premises that are provided with only a single stairway from upper floors or basements storeys are not normally suitable for more than 60 people. Similarly, if the ground floor of any part of your premises has only a single final exit it should be limited to no more than 60 people.

Ground and up to three upper storeys

Because of the higher risk posed by residential care premises with a single stairway to parts of the premises, and to protect the escape route by preventing smoke from entering the stairway, a protected lobby or corridor approach between the stairway and all floors (other than the top floor) should be provided as shown.

If your premises have a ground floor and up to three upper storeys and are served by a single stairway, it is important to understand that you are unlikely to be able to meet the suggested travel distance to a final exit (see Table 2 on page 76). Therefore the stairway needs to be protected by a fire-resisting enclosure as shown.

The layout shown in Figure 46 will be generally acceptable as long as the following apply:

- The furthest point on all of your floors to the lobbied storey exit is within the overall suggested travel distance (see Table 2).

- When a protected lobby or corridor approach to the stairway is employed the travel distance is measured to the storey exit and not the door to the lobby or corridor.

- Where the building incorporates a basement, any stairway from the basement is separated by a protected lobby or corridor between that basement and the protected stairway.

Exceptionally, in non sleeping areas with very low risk (where densities are low, occupants are familiar with the premises, excellent visual awareness and very limited combustibles) automatic fire detection on all floors may be used instead of protected lobbies or corridors (see Figure 47), however, the stairway must still be protected.

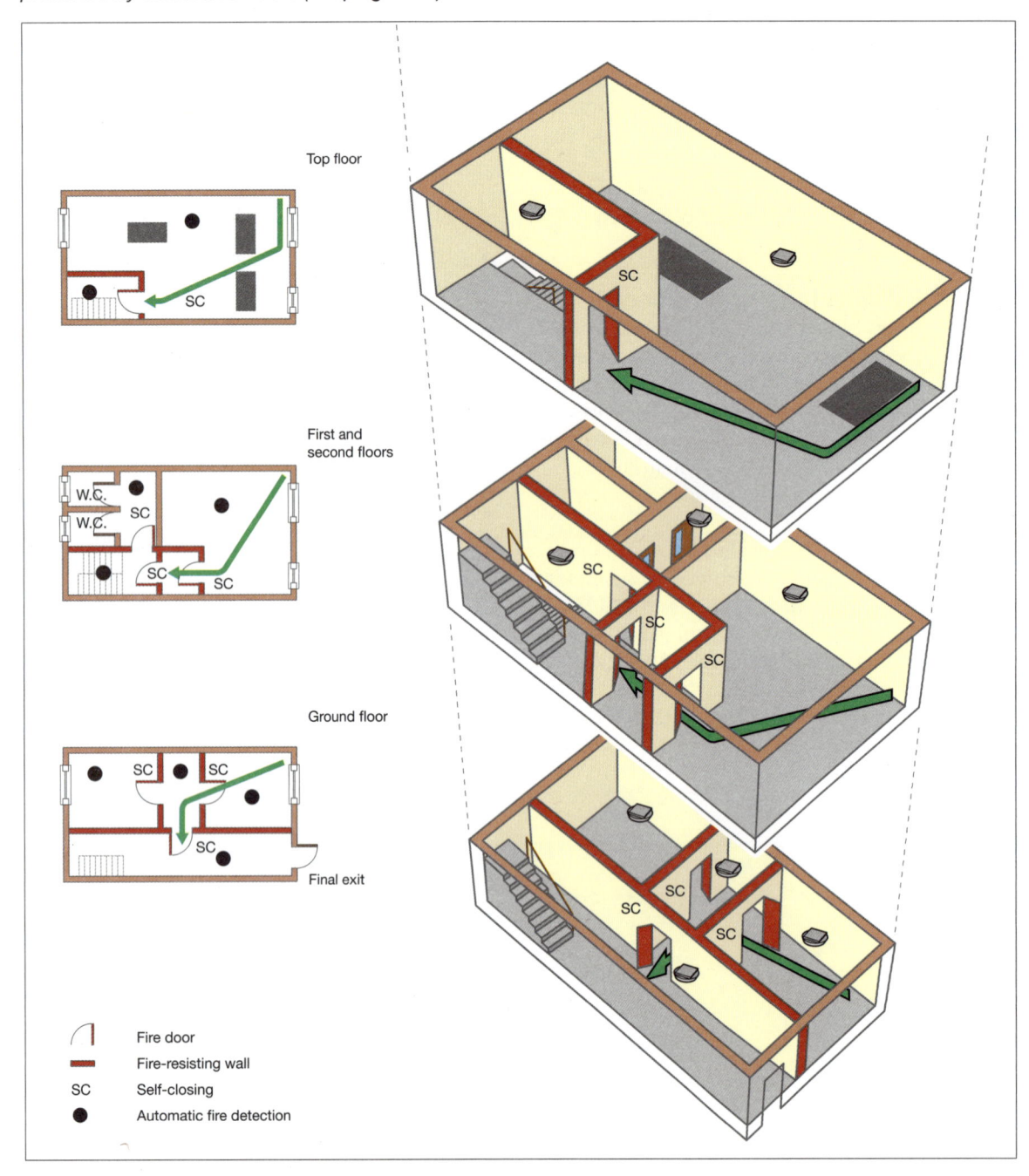

Figure 46: Ground and up to three upper storeys – protected by lobbies/corridors (sleeping areas)

Figure 47: Ground and up to three upper storeys – protected with automatic fire detection (non-sleeping areas with very low risk)

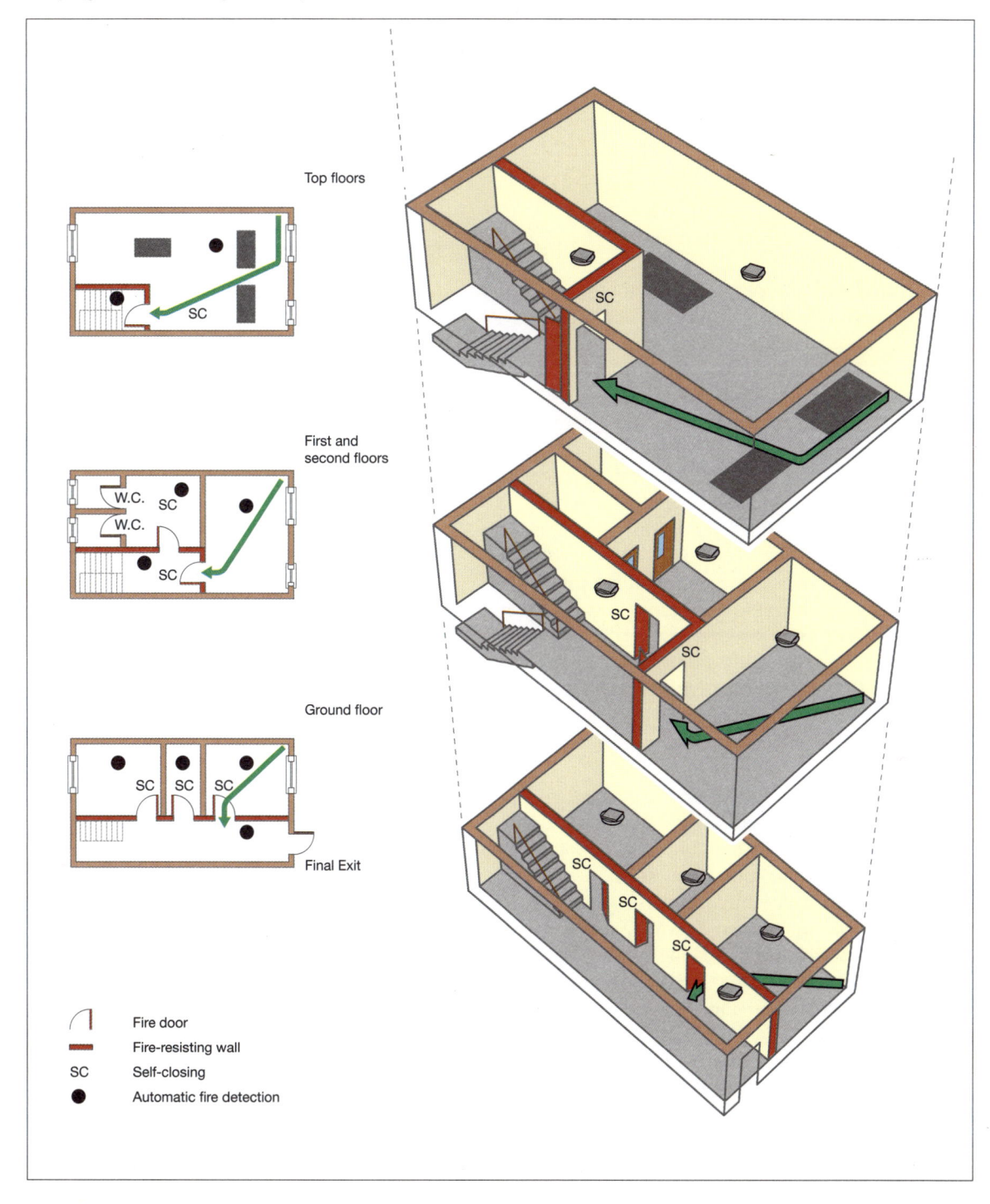

Section 5 Further guidance on emergency escape route lighting

The primary purpose of emergency escape lighting is to illuminate escape routes, but it also illuminates other safety equipment.

The size and type of your premises and the risk to the occupants will determine the complexity of the emergency escape lighting required.

In small premises where borrowed lighting is not appropriate single 'stand-alone' emergency escape lighting units may be sufficient and these can sometimes be combined with exit or directional signs. The level of general illumination should not be significantly reduced by the sign.

For most care homes, a more comprehensive system of fixed automatic emergency escape lighting is likely to be needed, especially in larger, more complex multi-storey premises, particularly in those with extensive basements or where there are significant numbers of residents, staff or members of the public.

You will have identified the escape routes when carrying out your fire risk assessment and need to ensure that they are all adequately lit. If there are escape routes that are not permanently illuminated by normal lighting, such as external stairs, then a switch, clearly marked 'Escape lighting', or some other means of switching on the lighting should be provided at the entry to that area/stairs.

An escape lighting system provided for escape purposes would normally cover the following:

- each exit door;
- escape routes;
- intersections of corridors;
- outside each final exit and on external escape routes;
- emergency escape signs;
- stairways so that each flight receives adequate light;
- changes in floor level;
- windowless rooms and toilet accommodation exceeding 8m^2;
- firefighting equipment;
- fire alarm call points;
- equipment that would need to be shut down in an emergency;
- lifts; and
- halls or other areas greater than 60m^2.

It is not necessary to provide individual lights (luminaires) for each item above, but there should be a sufficient overall level of light to allow them to be visible and usable.

Emergency escape lighting can be both 'maintained', i.e. on all the time, or 'non-maintained' which only operates when the normal lighting fails. Systems or individual units are available with durations of between one and three hours. In practice, the three-hour design is the most popular and can help with maintaining limited continued use of your premises during a power failure (other than in an emergency situation).

Figure 48: Luminaires

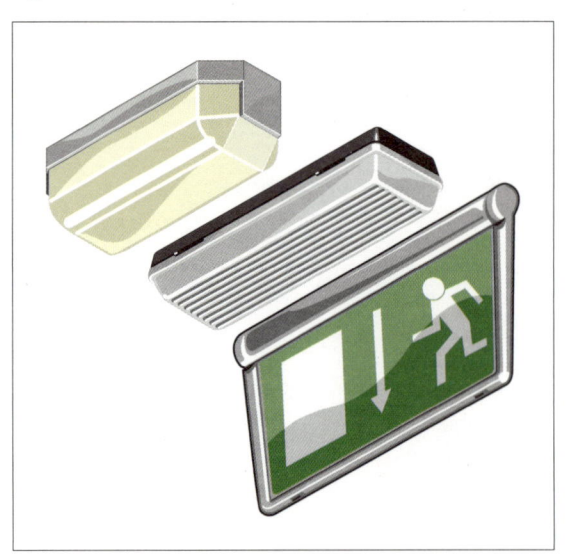

Emergency escape lighting luminaires can be stand-alone dedicated units or incorporated into normal light fittings. There are highly decorative versions of these for those areas that demand aesthetically pleasing fixtures. Power supplies can be rechargeable batteries integral to each unit, a central battery bank or an automatic start generator.

To complement emergency escape lighting, you may wish to identify exit routes by the use of way-guidance equipment. Way-guidance systems usually comprise photo-luminescent material, lines of LEDs, or strips of miniature incandescent lamps, forming a continuous marked escape route at lower level (see Figure 49). These systems have proved particularly effective when people have to escape through smoke, and for partially sighted people. They can be particularly useful in premises providing residential care where they can provide marked routes on floors and in multi-storey premises they can direct people to escape routes which are seldom used.

If you decide that you need to install emergency escape lighting or to modify your existing system, any work should be carried out by a competent person in accordance with the appropriate standards. Further guidance is given in BS 5266-1[27A] and BS EN 50172 (BS 5266-8).[27]

Figure 49: A 'way-guidance' system

Maintenance and testing of emergency escape lighting

All emergency escape lighting systems should be regularly tested and properly maintained to an appropriate standard. Most existing systems will need to be manually tested. However, some modern systems have self-testing facilities that reduce routine checks to a minimum.

Depending on your type of installation you should be able to carry out most of the routine tests yourself. The test method will vary. If you are not sure how to carry out these tests you should contact your supplier or other competent person.

Figure 50: A test key

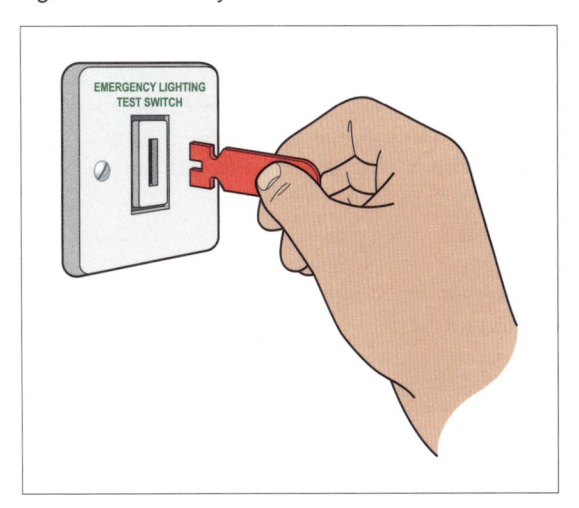

Test facilities often take the form of a 'fishtail' key (see Figure 50) inserted in a special switch either near the main fuse board or adjacent to relevant light switches.

Typically, testing would include:

- a daily visual check of any central controls;
- a monthly function test by operating the test facility for a period sufficient to ensure that each emergency lamp illuminates; and
- an annual full discharge test.

Particular care needs to be taken following a full discharge test. Batteries typically take 24 hours to re-charge and the premises should not be occupied until the emergency lighting system is fully functioning unless alternative arrangements have been made. This will be of particular importance in premises providing sleeping accommodation seven days a week. Further guidance on testing is given in BS EN 50172 (BS 5266-8).[27]

It is good practice to keep a record of tests.

Section 6 Further guidance on signs and notices

Escape signs

In simple premises, a few signs indicating the alternative exit(s) might be all that is needed. In larger and more complex premises, a series of signs directing people along the escape routes towards the final exit might be needed.

Many people with poor vision retain some sight and are able to recognise changing or contrasting colour to provide them with visual clues when moving around a building.* It may be sufficient to paint any columns in a contrasting colour (see BS 8300[14]) and to highlight changes in level by, for example, making the nosing to step and stair treads a contrasting colour.

For people with no sight, a well-managed 'buddy system', continuous handrails or a sound localisation system (which helps people to move towards an alert sound), or the installation of more tactile aids may be appropriate.

People will usually attempt to leave premises by the same way that they entered. As these entrances in places of residential care will then be the exits of choice it follows that the other exit facilities provided for the premises may not be fully utilised. For this reason all available exits should be clearly indicated so that residents and the public are aware that there are ways to leave the building other than by the doors which they used to gain admission. In addition, the provision of well sign-posted exits in full view of an audience can give a feeling of security in an emergency situation.

Maintained internally illuminated exit signs will be necessary where the lighting may be dimmed or extinguished (e.g. rest rooms).

Exit signs should be clearly visible whenever residents, the public, staff and contractors are present.

Positioning of escape route signs

The presence of other signs in homes (such as notices and residents' information) or decorations, can distract attention from, or obscure the visibility of escape signs. This could affect people's ability to see and understand escape signs, particularly if there is a fire evacuation. Always ensure that escape signs are not overwhelmed.

Escape signs should meet the following criteria:

- They should provide clear, unambiguous information to enable people to leave a building safely in an emergency.

- Every escape route sign should, where necessary, incorporate, or be accompanied by, a directional arrow. Arrows should not be used on their own.

- If the escape route to the nearest exit is not obvious then it should be indicated by a sign(s).

- They should be positioned so that a person escaping will always have the next escape route sign in sight.

- They should be fixed above the door in the direction of escape and not be fixed to doors, as they will not be visible if the door is open.

- Signs mounted above doors should be at a height of between 2.0m and 2.5m above the floor.

- Signs on walls should be mounted between 1.7m and 2.0m above the floor.

- Mounting heights greater than 2.5m may be used for hanging signs, e.g. in large open spaces or for operational reasons, but care should be taken to ensure that such signs are both conspicuous and legible. In such cases larger signs may be necessary.

- Signs should be sited at the same height throughout the escape route, as far as is reasonably practicable.

---

* The Royal National Institute of the Blind estimates that only about 4% of visually impaired people are totally blind.

Escape sign design
For a sign to comply with safety signs and signals regulations it must be pictographic (see Figures 51 and 52). The pictogram can be supplemented by text if this is considered necessary to make the sign easily understood (BS-type sign) but you must not have a fire safety sign that uses only text. Either type of sign can be used but different types should not be mixed. Appropriate signs should take into account the needs of those who may need to use them.

Figure 51: BS-type sign

Figure 52: Euro sign

The legibility of escape signs is determined by the size of the sign, its level of illumination and the distance over which it is viewed. The use of signs within the same premises should follow a consistent design pattern or scheme. You should not rely on a few outsized signs which may encourage people to travel to a particular escape route when other more appropriate routes should be used.

Other safety signs and notices
A number of other mandatory signs such as 'Fire action' notices may also be necessary.

Fire doors that have been fitted with self-closing devices should be labelled 'Fire door – keep shut' on both sides. Fire-resisting doors to cupboards, stores and service ducts that are not self-closing because they are routinely kept locked should be labelled 'Fire door – keep locked' on the outside.

Figure 53: 'Fire door keep shut' notice

Signs should indicate non-automatic fire safety equipment if there is any doubt about its location, e.g. fire extinguishers that are kept in cabinets or in recesses.

A notice with the words 'Push bar to open' should be permanently displayed immediately above the push-bar on all doors fitted with a panic bolt or panic latch.

A notice with the words 'Fire escape – Keep clear' should be permanently displayed at about eye level on the external face of all doors which are provided as a means of escape in case of fire and which, because they are not normally used, may become obstructed.

Staff notices
In simple premises providing residential care where there are a limited number of people, it may be reasonable to provide staff with verbal reminders of what they need to do if there is a fire. You could consider providing this in a short written statement that could, for example, be delivered with staff pay slips every six months.

In larger and more complex premises or where there is a high turnover of staff, a more considered approach for staff notices and instructions will be necessary. As well as positioning the fire instructions notice on escape routes adjacent to fire break-glass call points, put them where staff frequently assemble in the premises, e.g. the staff room and locker rooms.

Figure 54: A typical staff action notice

If you have staff whose first language is not English you may need to consider providing instructions in more than one language. The interpretation should always convey an identical message.

Illumination
All signs and notices will need illumination to ensure they are conspicuous and legible. Options available to achieve this include:

- external illumination; and
- internal illumination.

The supplier or other competent person can give you further advice.

Signs or notices of the photo-luminescent type, i.e. where the active material making up the luminous parts of such signs or notices needs a period of exposure to light before they become visible in darkness (but get fainter with time), are not a substitute for appropriate emergency lighting and should only be used where other forms of illumination are present.

Further guidance
Detailed guidance on fire safety signs can be found in BS 5499-4[90] and BS 5499-5.[29] Published guidance[5,6] on compliance with health and safety legislation on signs is also available. Guidance on photoluminescent fire safety signs and notices is given in BS 5266-6.[91]

Section 7 Further guidance on recording, planning, informing, instructing and training

7.1 Fire safety records

Keeping up-to-date records of your fire risk assessment can help you effectively manage the fire strategy for your premises and demonstrate how you are complying with fire safety law.

Even if you do not have to record the fire risk assessment, it can be helpful to keep a record of any co-operation and exchange of information made between managers and other responsible people for future reference.

In larger and more complex premises, it is best to keep a dedicated record of all maintenance of fire protection equipment and training. There is no 'correct' format specified for this. Suitable record books are available from trade associations and may also be available from your local enforcing authority.

In all cases the quality of records may also be regarded as a good indicator of the overall quality of the safety management structure.

Fire safety records could include:

- details of any significant findings from the fire risk assessment and any action taken (see Part 1, Section 4.1);
- testing and checking of escape routes, including final exit locking mechanisms, such as panic devices, emergency exit devices and any electromagnetic devices;
- testing of fire warning systems, including weekly alarm tests and periodic maintenance by a competent person;
- recording of false alarms;
- testing and maintenance of emergency lighting systems;
- testing and maintenance of fire extinguishers, hose reels and fire blankets etc.;
- if appropriate, testing and maintenance of other fire safety equipment such as fire-suppression and smoke control systems;
- recording and training of relevant people and fire evacuation drills;
- planning, organising, policy and implementation, monitoring, audit and review;
- maintenance and audit of any systems that are provided to help the fire and rescue service;
- the arrangements for a co-ordinated emergency plan or overall control of the actions you or your staff should take if there is a fire;
- all alterations, tests, repairs and maintenance of fire safety systems, including passive systems such as fire doors; and
- all 'near miss' events and 'lessons learned', and any remedial action taken.

Other issues that you may wish to record include:

- the competence, qualifications and status of the persons responsible for carrying out inspections and tests;
- the results of periodic safety audits, reviews, inspections and tests, and any remedial action taken;
- all incidents and circumstances which had the potential to cause accidents and monitor subsequent remedial actions; and
- a record of the building use, the fire prevention and protection measures in place and high-risk areas.

You should ensure that no other management decisions or policies compromise safety.

Your documentation should be available for inspection by representatives of the enforcing authority.

More detailed advice is given in BS5588-12.[75]

Figure 55 is an example of how to record some individual stages of the process in more detail.

Figure 55: Example record of significant findings

Risk Assessment – Record of significant findings				
Risk assessment for			**Assessment undertaken by**	
Company	Happy Valley Care Home		Date	10/03/2006
Address	Green Lane Anytown AA6 3ZS		Completed by Signature	J Brown
Sheet number		**Floor/area**		**Use**
One		Ground		Day room, kitchen, laundry and accommodation

Step 1 – Identify fire hazards

Sources of ignition	Sources of fuel	Sources of oxygen
• Heat from cooking • Smoking materials • Heat from open fire	• Cooking oil • Upholstered furniture/bedding/laundry • Clothing	• No additional sources

Step 2 – People at risk

- Staff in kitchen, residents in dining room and accommodation on all floors
- Staff and residents on all floors
- Residents in lounge

Step 3 – Evaluate, remove, reduce and protect from risk

(3.1) Evaluate the risk of the fire occuring	• Risk of ignition of hot oil and spread to extractor • High risk in resident bedrooms where smoking occurs, low elsewhere • Medium risk of ignition to residents clothing
(3.2) Evaluate the risk to people from a fire starting in the premises	• Fire in kitchen would be detected quickly but could spread to dining room • Fire in any part would be detected quickly but risk to residents in room or origin • Any fire likely to be local and cause little risk to others
(3.3) Remove and reduce the hazards that may cause a fire	• Unable to remove • Unable to remove totally. Ensure all residents are aware of the dangers • Regularly check size of fire and ensure guard always in place
(3.4) Remove and reduce the risks to people from a fire	• Train staff in safe cooking practices and use of extinguishers • Ensure all soft furniture and bedding is fire retardent • Patrol regularly to ensure residents remain at a safe distance from fire

Assessment review

Assessment review date	Completed by	Signature

Review outcome (where substantial changes have occurred a new record sheet should be used)

Notes:
(1) The risk assessment record of significant findings should refer to other plans, records or other documents as necessary.
(2) The information in this record should assist you to develop an emergency plan; coordinate measures with other 'responsible persons' in the building; and to inform and train staff and inform other relevant persons.

In premises with 'engineered fire safety strategies', a fire policy manual should be provided in addition to any other records. Enforcing authorities would expect a fire engineering policy manual to conform to an acceptable structure. BS 7974-0 Section 5: Reporting and presentation gives further guidance.[30]

Fire safety audit

A fire safety audit can be used alongside your fire risk assessment to identify what fire safety provisions exist in your premises.

When carrying out a review of your fire safety risk assessment, a pre-planned audit can quickly identify if there have been any significant changes which may affect the fire safety systems and highlight whether a full fire risk assessment is necessary.

Plans and specifications

Plans and specifications can assist understanding of a fire risk assessment or emergency plan. Even where not needed for this purpose they can help you and your staff keep your fire risk assessment and emergency plan under review and help the fire and rescue service in the event of fire. Any symbols used should be shown on a key. Plans and specifications could include the following:

- essential structural features such as the layout of the sleeping areas and other function rooms, escape doors, wall partitions, corridors, stairways, etc. (including any fire-resisting structure and self-closing doors provided to protect the escape route);
- location of protected areas that have been designated suitable for the horizontal evacuation of mobility-impaired people;
- location of refuges and lifts that have been designated suitable for use by disabled people and others who may need assistance to escape in case of a fire;
- methods for fighting fire (details of the number, type and location of the firefighting equipment);
- location of manually-operated fire alarm call points and control equipment for fire alarms;
- location of any emergency lighting equipment and the exit route signs;
- location of any equipment or process that must be immediately shut down by staff on hearing the fire alarm;
- location of any automatic firefighting systems and sprinkler control valves;
- location of the main electrical supply switch, the main water shut-off valve and, where appropriate, the main gas or oil shut-off valves; and
- plans and specifications relating to all recent constructions.

This information should be passed on to any later users or owners of the premises.

7.2 Emergency plans

Emergency plan and contingency plan

Your emergency plan should be appropriate to your premises and could include:

- how people will be warned if there is a fire;
- what staff should do if they discover a fire;
- how the evacuation of the premises should be carried out;
- individual needs/risks relating to individual residents;
- identification and use of protected areas for horizontal evacuation;
- procedures for a 'delayed evacuation' response;
- where people should assemble after they have left the premises and procedures for checking whether the premises have been evacuated;
- identification of key escape routes, how people can gain access to them and escape from them to a place of total safety;
- arrangements for fighting fire;
- the duties and identity of staff who have specific responsibilities if there is a fire;
- arrangements for the safe evacuation of people identified as being especially at risk, such as residents and others with disabilities, children and people working alone;
- any machines/processes/appliances/power supplies that need to be stopped or isolated if there is a fire;
- specific arrangements, if necessary, for high-fire-risk areas;

- contingency plans, such as restrictions on the use of the building) for when life safety systems, such as evacuation lifts, fire-detection and warning systems, sprinklers or smoke control systems are out of order;

- how the fire and rescue service and any other necessary services will be called and who will be responsible for doing this; and

- procedures for meeting the fire and rescue service on their arrival and notifying them of the locations of any remaining residents and of any special risks, e.g. the location of highly flammable materials.

It can be helpful to keep an 'emergency box' containing up to date information on all the residents present, including their medication, special needs and next of kin. A member of staff should be charged with taking the box out of the premises when the fire warning actuates, and, where appropriate, making information from it available to the emergency services.

As part of your emergency plan it is good practice to prepare post-incident plans for dealing with situations that might arrise, such as those involving:

- people with personal belongings (especially valuables) still in the building;
- people in a state of undress;
- getting people away from the building (e.g. to transport);
- arranging alternative accommodation; and
- inclement weather.

You should also prepare contingency plans to determine specific actions and or the mobilisation of specialist resources.

Guidance on developing health and safety management policy has been published by the HSE.[31]

7.3 Information, instruction, co-operation and co-ordination

Supplying information
You must provide easily understandable information to employees, the parents of children you may employ, and to employers of other persons working in your premises about the measures in place to ensure a safe escape from the building and how they will operate, for example:

- any significant risks to staff, residents and other relevant persons that have been identified in your fire risk assessment or any similar assessment carried out by another user and responsible person in the building;

- the fire prevention and protection measures and procedures in your premises, and where they impact on staff and other relevant persons in the building;

- the procedures for fighting a fire in the premises; and

- the identity of people who have been nominated with specific responsibilities in the building.

Even if you do not have to record the fire risk assessment, it would be helpful to keep a record of any cooperation and exchange of information made between employers and other responsible people for future reference.

You need to ensure that all staff and, where necessary, other relevant persons who work in the building, receive appropriate information given in a way that can be easily understood. This might include any special instructions to particular people who have been allocated a specific task, such as shutting down equipment or guiding residents and visitors to the nearest exit.

Duties of employees to give information
Employees also have a duty to take reasonable care for their own safety and that of other people who may be affected by their activities. This includes the need for them to inform their employer of any activity that they consider would present a serious and immediate danger to their own safety and that of others.

Dangerous substances
HSE publishes guidance[8] about specific substances where appropriate information may need to be provided. If any of these, or any other substance that is not included but nevertheless presents more than a slight risk, is present in premises, then you must provide such information to staff and others. Specifically, you must:

- name of the substance and the risks associated with it, e.g. how to safely use or store the product to avoid creating highly flammable vapoures or explosive atmospheres (to include the use of bottled or piped oxygen, and aerosols);

- identify any legislative provisions that may be associated with the substance;

- allow employees access to the hazardous substances safety data sheet; and
- inform the local fire and rescue service where dangerous substances are present on the premises.

Information to the fire and rescue service

In addition to providing information to the fire and rescue service when dangerous substances are present in sufficient quantities to pose an enhanced risk, it will also be helpful to inform them of any short term changes that might have an impact on their firefighting activities; e.g. in the event of temporary loss of a firefighting facility and temporary alterations. Procedures should also include meeting and briefing the fire and rescue service when they arrive.

Case study

If the access roads to a care premises become unusable, e.g. because of building work, this should be brought to the attention of the fire and rescue service. Being unable to use access roads to allow fire engines up to the building might have a serious effect on the ability of firefighters to begin operations as quickly as planned. The information supplied will enable the emergency services to make adjustments to the level of the emergency response.

Instruction

You will need to carefully consider the type of instructions to staff and other people working in your premises. Written instructions must be concise, comprehensible and relevant, and therefore must be reviewed and updated as new working practices and hazardous substances are introduced.

Inclusive access and employment policies mean that people with learning difficulties may now be present in a range of premises and your fire risk assessment should consider whether further instruction or guidance is necessary to ensure that your evacuation strategy is appropriate and understood by everyone.

Instructions will need to be given to people delegated to carry out particular tasks, for example:

- checking that escape routes are accessible;
- daily, weekly, quarterly and yearly checks on the range of fire safety measures (in larger premises some of the work may be contracted out to a specialist company);
- safety considerations at the end of each day, e.g. removing rubbish, ensuring exits are usable and closing fire doors;
- ensuring that hazardous substances are in a safe condition;
- the safe storage of hazardous substances at the end of the working day; and
- ensuring everyone knows how to use internal emergency telephones.

Specific instruction may be needed about:

- how staff will assist residents and visitors to evacuate the premises in line with pre-determined plans;
- 'sweeping' each floor by staff to guide people to the nearest exit when the fire alarm sounds;
- designating particular areas of the premises for staff to check that no one remains inside;
- calling the emergency services when a fire is confirmed or cannot be ruled out;
- carrying out evacuation roll calls;
- taking charge at the assembly area;
- meeting and directing fire engines; and
- cover arrangements when nominated people are on leave.

Co-operation and co-ordination

Where you share premises with others (e.g. as part of a complex), each responsible person, i.e. each employer, owner or other person who has control over any part of the premises, will need to co-operate and co-ordinate the findings of their separate fire risk assessments to ensure the fire precautions and protection measures are effective throughout the building. This could include:

- co-ordinating an emergency plan (see Step 4.2 for features of an emergency plan);
- identifying the nature of any risks and how they may affect others in or about the complex;
- identifying any fire prevention and protection measures;
- identifying any measures to mitigate the effects of a fire; and

- arranging any contacts with external emergency services and calling the fire and rescue services.

7.4 Fire safety training

Staff training

The actions of staff if there is a fire are likely to be crucial to their safety and that of other people in the premises. All staff should receive basic fire safety induction training and attend refresher sessions at pre-determined intervals.

You should ensure that all staff and contractors (and, if appropriate, residents and regular visitors) are told about the emergency plan and are shown the escape routes.

The training should take account of the findings of the fire risk assessment and be easily understood by all those attending. It should include the role that those members of staff will be expected to carry out if a fire occurs. This may vary in large premises, with some staff being appointed as fire marshals or being given some other particular role for which additional training will be required.

In addition to the guidance given in Part 1, Step 4.4, as a minimum all staff should receive training about:

- the items listed in your emergency plan;
- the importance of fire doors and other basic fire prevention measures;
- where relevant, the appropriate use of firefighting equipment;
- the importance of reporting to the assembly area;
- exit routes and the operation of exit devices, including physically walking these routes; and
- general matters such as permitted smoking areas or restrictions on cooking other than in designated areas.

Training is necessary:

- when staff start employment or are transferred into the premises;
- when changes have been made to the emergency plan and the preventive and protective measures;
- where working practices and processes or people's responsibilities change;

- to take account of any changed risks to the safety of staff or other relevant persons; and
- to ensure that staff know what they have to do to safeguard themselves and others on the premises.

In situations where residents of poor mobility are to be physically moved by staff, you must ensure that the staff have received the appropriate training to do so. If evacuation aids or equipment is provided for this purpose, staff should be trained to deploy and utilise them as quickly and efficiently, and with as little distress to the resident as possible.

Training should be repeated as often as necessary and should take place during the working hours of individual staff members.

Whatever training you decide is necessary to support your fire safety strategy and emergency plan, it should be verifiable.

Enforcing authorities may want to examine records as evidence that adequate training has been given.

Fire marshals

Staff expected to undertake the role of fire marshals (often called fire wardens) would require more comprehensive training. Their role may include:

- helping those on the premises to leave;
- checking the premises to ensure everyone has left;
- using firefighting equipment if safe to do so;
- liaising with the fire and rescue service on arrival;
- shutting down vital or dangerous equipment; and
- performing a supervisory/managing role in any fire situation.

Training for this role may include:

- detailed knowledge of the fire safety strategy of the premises;
- awareness of human behaviour in fires;
- how to encourage others to use the most appropriate escape route;
- how to search safely and recognise areas that are unsafe to enter;

- the difficulties that some people, particularly if disabled, may have in escaping and any special evacuation arrangements that have been pre-planned;
- additional training in the use of firefighting equipment;
- an understanding of the purpose of any fixed firefighting equipment such as sprinklers or gas flooding systems; and
- reporting of faults, incidents and 'near misses'.

Fire drills

Once the emergency plan has been developed and training given, you will need to evaluate its effectiveness. The best way to do this is to perform a fire drill. This should be carried out at least annually or as determined by your fire risk assessment. If you have a high staff turnover, you may need to carry them out more often.

A well-planned and executed fire drill will confirm understanding of the training and provide helpful information for future training. The responsible person should determine the possible objectives of the drill such as to:

- identify any weaknesses in the evacuation strategy;
- test the procedure following any recent alteration or changes to working practices;
- test the procedure with minimum staff levels;
- familiarise new members of staff with procedures;
- familiarise residents with procedures; and
- test the arrangements for disabled people;

Who should take part?

Within each building the evacuation should be for all occupants except those who may need to ensure the security of the premises, or people who, on a risk-assessed basis, cannot be moved, must remain with a resident who cannot be moved, or are required to remain with particular equipment or processes that cannot be closed down.

Premises or complexes that consist of several buildings on the same site should be dealt with one building at a time over an appropriate period unless the emergency procedure dictates otherwise.

You should consider the feasibility of drills involving non-ambulant or semi-ambulant residents where their medical condition permits. Properly simulated drills will include residents, or a significant number of residents. It is advisable to either have one drill at night or to simulate night time conditions, which would include the applicable night staffing levels.

The frequency of drills should ensure that regular occasional residents (e.g. who attend the home for a few days each month for respite care) are included in some drills.

You may find it helpful to include members of the public (visitors) in your fire drill – ensuring that all necessary health and safety issues are addressed before you do so.

Carrying out the drill

For premises that have more than one escape route, the escape plan should be designed to evacuate all people on the assumption that one exit or stairway is unavailable because of the fire. This could be simulated by a designated person being located at a suitable point on an exit route. Applying this scenario to different escape routes at each fire drill will encourage individuals to use alternative escape routes which they may not normally use.

When carrying out the drill you might find it helpful to:

- circulate details concerning the drill and inform all staff of their duty to participate. It may not be beneficial to have 'surprise drills' as the health and safety risks introduced may outweigh the benefits;
- inform residents with learning difficulties by using special techniques (e.g. visual imagery);
- ensure that equipment can be safely left;
- nominate observers;
- inform the alarm receiving centre if the fire warning system is monitored (if the fire and rescue service is normally called directly from your premises, ensure that this does not happen);
- inform visitors if they are present; and
- ask a member of staff at random to set off the alarm by operating the nearest alarm call point using the test key. This will indicate the level of knowledge regarding the location of the nearest call point (a different call point should be used on different drills).

More detailed information on fire drills and test evacuations are given in BS5588-12.[75]

The roll call/checking the premises have been evacuated

Carry out a roll call as soon as possible at the designated assembly point(s), and/or receive reports from wardens designated to 'sweep' the premises. You should note any people who are unaccounted for. In a real evacuation this information will need to be passed to the fire and rescue service on arrival.

Check that people have assembled at the evacuation point

Once the roll call is complete or all reports have been received, allow people to return to the building. If the fire warning system is monitored inform the alarm receiving centre that the drill has now been completed and record the outcomes of the drill.

In premises where residents may not be fully aware of what is happening (due to age, or mental impairment) then staff will have to act on their behalf. It is also possible that some residents will wander away from the assembly point and this should be monitored, and if possible controlled.

Monitoring and debrief

Throughout the drill the responsible person and nominated observers should pay particular attention to:

- difficulties in moving, or the time taken to move, non- or semi-ambulant residents;

- communication difficulties with regard to the roll call and establishing that everyone is accounted for;

- the use of the nearest available escape routes as opposed to common circulation routes;

- difficulties with the opening of final exit doors;

- difficulties experienced by people with disabilities;

- the roles of specified people, e.g. fire wardens;

- inappropriate actions, e.g. stopping to collect personal items, attempting to use lifts, etc.; and

- windows and doors not being closed as people leave.

On-the-spot debriefs are useful to discuss the fire drill, encouraging feedback from everybody. Later, reports from fire wardens and observations from people should be collated and reviewed. Any conclusions and remedial actions should be recorded and implemented.

Section 8 Quality assurance of fire protection equipment and installation

Fire protection products and related services should be fit for their purpose and properly installed and maintained in accordance with the manufacturer's instructions or a relevant standard.

Third party certification schemes for fire protection products and related services are an effective means of providing the fullest possible assurances, offering a level of quality, reliability and safety that non-certificated products may lack. This does not mean goods and services that are not third party approved are less reliable, but there is no obvious way in which this can be demonstrated.

Third party quality assurance can offer comfort, both as a means of satisfying you that goods and services you have purchased are fit for purpose, and as a means of demonstrating that you have complied with the law.

However, to ensure the level of assurance offered by third party schemes, you should always check whether the company you employ sub-contracts work to others. If they do, you will want to check that the sub-contractors are subject to the same level of checks of quality and competence as the company you are employing.

Your local fire and rescue service, fire trade associations or your own trade association may be able to provide further details about third party quality assurance schemes and the various organisations that administer them.

Appendix A

A.1 Example fire safety maintenance checklist

A fire safety maintenance checklist can be used as a means of supporting your fire safety policy. This example list is not intended to be comprehensive and should not be used as a substitute for carrying out a fire risk assessment.

You can modify the example, where necessary, to fit your premises and may need to incorporate the recommendations of manufacturers and installers of the fire safety equipment/systems that you may have installed in your premises.

Any ticks in the grey boxes should result in further investigation and appropriate action as necessary. In larger and more complex premises you may need to seek the assistance of a competent person to carry out some of the checks.

	Yes	No	N/A	Comments
Daily checks (not normally recorded)				
Escape routes				
Can all fire exits be opened immediately and easily?	☐	☐	☐	
Are fire doors clear of obstructions?	☐	☐	☐	
Are escape routes clear?	☐	☐	☐	
Fire warning systems				
Is the indicator panel showing 'normal'?	☐	☐	☐	
Are whistles, gongs or air horns in place?	☐	☐	☐	
Escape lighting				
Are luminaires and exit signs in good condition and undamaged?	☐	☐	☐	
Is emergency lighting and sign lighting working correctly?	☐	☐	☐	
Firefighting equipment				
Are all fire extinguishers in place?	☐	☐	☐	
Are fire extinguishers clearly visible?	☐	☐	☐	
Are vehicles blocking fire hydrants or access to them?	☐	☐	☐	
Weekly checks				
Escape routes				
Do all emergency fastening devices to fire exits (push bars and pads, etc.) work correctly?	☐	☐	☐	
Are external routes clear and safe?	☐	☐	☐	
Fire warning systems				
Does testing a manual call point send a signal to the indicator panel? (Disconnect the link to the receiving centre or tell them you are doing a test.)	☐	☐	☐	
Did the alarm system work correctly when tested?	☐	☐	☐	
Did staff and other people hear the fire alarm?	☐	☐	☐	
Did any linked fire protection systems operate correctly? (e.g. magnetic door holder released, smoke curtains drop)	☐	☐	☐	

	Yes	No	N/A	Comments
Weekly checks *continued*				
Do all visual alarms and/or vibrating alarms and pagers (as applicable) work?	☐	☐	☐	
Do voice alarm systems work correctly? Was the message understood?	☐	☐	☐	
Escape lighting				
Are charging indicators (if fitted) visible?	☐	☐	☐	
Firefighting equipment				
Is all equipment in good condition?	☐	☐	☐	
Additional items from manufacturer's recommendations.	☐	☐	☐	
Monthly checks				
Escape routes				
Do all electronic release mechanisms on escape doors work correctly? Do they 'fail safe' in the open position?	☐	☐	☐	
Do all automatic opening doors on escape routes 'fail safe' in the open position?	☐	☐	☐	
Are fire door seals and self-closing devices in good condition?	☐	☐	☐	
Do all roller shutters provided for fire compartmentation work correctly?	☐	☐	☐	
Are external escape stairs safe?	☐	☐	☐	
Do all internal self-closing fire doors work correctly?	☐	☐	☐	
Escape lighting				
Do all luminaires and exit signs function correctly when tested?	☐	☐	☐	
Have all emergency generators been tested? (Normally run for one hour.)	☐	☐	☐	
Firefighting equipment				
Is the pressure in 'stored pressure' fire extinguishers correct?	☐	☐	☐	
Additional items from manufacturer's recommendations.	☐	☐	☐	
Three-monthly checks				
General				
Are any emergency water tanks/ponds at their normal capacity?	☐	☐	☐	
Are vehicles blocking fire hydrants or access to them?	☐	☐	☐	
Additional items from manufacturer's recommendations.	☐	☐	☐	
Six-monthly checks				
General				
Has any firefighting or emergency evacuation lift been tested by a competent person?	☐	☐	☐	
Has any sprinkler system been tested by a competent person?	☐	☐	☐	
Have the release and closing mechanisms of any fire-resisting compartment doors and shutters been tested by a competent person?	☐	☐	☐	
Fire warning system				
Has the system been checked by a competent person?	☐	☐	☐	

	Yes	No	N/A	Comments
Six-monthly checks *continued*				
Escape lighting				
Do all luminaires operate on test for one third of their rated value?	☐	☐	☐	
Additional items from manufacturer's recommendations.	☐	☐	☐	
Annual checks				
Escape routes				
Do all self-closing fire doors fit correctly?	☐	☐	☐	
Is escape route compartmentation in good repair?	☐	☐	☐	
Escape lighting				
Do all luminaires operate on test for their full rated duration?	☐	☐	☐	
Has the system been checked by a competent person?	☐	☐	☐	
Firefighting equipment				
Has all firefighting equipment been checked by a competent person?	☐	☐	☐	
Miscellaneous				
Has any dry/wet rising fire main been tested by a competent person?	☐	☐	☐	
Has the smoke and heat ventilation system been tested by a competent person?	☐	☐	☐	
Has external access for the fire service been checked for ongoing availability?	☐	☐	☐	
Have any firefighters' switches been tested?	☐	☐	☐	
Has the fire hydrant bypass flow valve control been tested by a competent person?	☐	☐	☐	
Are any necessary fire engine direction signs in place?	☐	☐	☐	

A2 Example form for recording significant findings

Risk Assessment – Record of significant findings		
Risk assessment for		**Assessment undertaken by**
Company		Date
Address		Completed by
		Signature
Sheet number	**Floor/area**	**Use**

Step 1 – Identify fire hazards

Sources of ignition	Sources of fuel	Sources of oxygen

Step 2 – People at risk

Step 3 – Evaluate, remove, reduce and protect from risk

(3.1) Evaluate the risk of the fire occuring

(3.2) Evaluate the risk to people from a fire starting in the premises

(3.3) Remove and reduce the hazards that may cause a fire

(3.4) Remove and reduce the risks to people from a fire

Assessment review

Assessment review date	Completed by	Signature

Review outcome (where substantial changes have occurred a new record sheet should be used)

Notes:
(1) The risk assessment record of significant findings should refer to other plans, records or other documents as necessary.
(2) The information in this record should assist you to develop an emergency plan; coordinate measures with other 'responsible persons' in the building; and to inform and train staff and inform other relevant persons.

Appendix B
Technical information on fire-resisting separation, fire doors and door fastenings

B1 Fire-resisting separation

General
The materials from which your premises are constructed may determine the speed with which a fire may spread, affecting the escape routes that people will use. A fire starting in a building constructed mainly from readily combustible material will spread faster than one where modern fire-resisting construction materials have been used. Where non-combustible materials are used and the internal partitions are made from fire-resisting materials, the fire will be contained for a longer period, allowing more time for the occupants to escape.

Because of the requirements of the Building Regulations you will probably already have some walls and floors that are fire-resisting and limitations on the surface finishes to certain walls and ceilings.

You will need to consider whether the standard of fire resistance and surface finishing in the escape routes is satisfactory, has been affected by wear and tear or alterations and whether any improvements are necessary.

The following paragraphs give basic information on how fire-resisting construction can provide up to 30 minutes protection to escape routes. This is the standard recommended for most situations. If you are still unsure of the level of fire resistance which is necessary after reading this information, you should consult a fire safety expert.

Fire-resisting construction
The fire resistance of a wall or floor is dependent on the quality of construction and materials used. Common examples of types of construction that provide 30-minute fire resistance to escape routes if constructed to the above standards are:

- internal framed construction wall, non-load bearing, consisting of 72mm x 37mm timber studs at 600mm centres and faced with 12.5mm of plasterboard with all joints taped and filled (see Figure 56);

- internal framed construction, non-load bearing, consisting of channel section steel studs at 600mm centres faced with 12.5mm of plasterboard with all joints taped and filled; and

- masonry cavity wall consisting of solid bricks of clay, brick earth, shale, concrete or calcium silicate, with a minimum thickness of 90mm on each leaf.

Figure 56: Fire-resisting construction

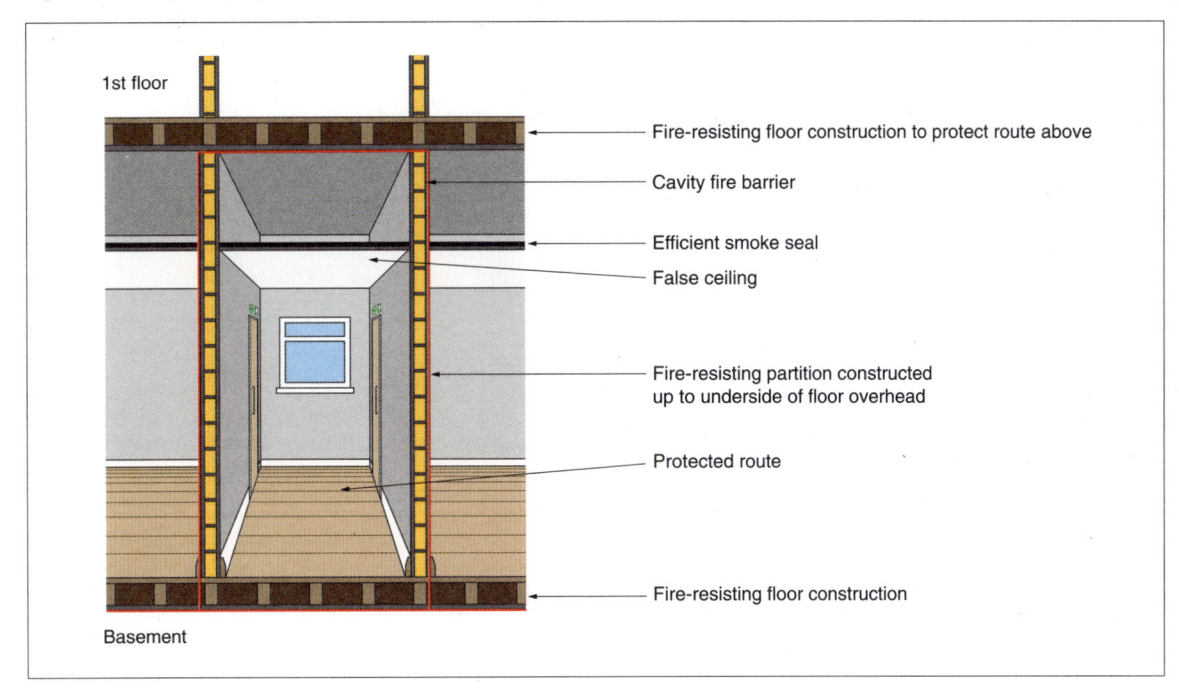

- 1st floor
- Fire-resisting floor construction to protect route above
- Cavity fire barrier
- Efficient smoke seal
- False ceiling
- Fire-resisting partition constructed up to underside of floor overhead
- Protected route
- Fire-resisting floor construction
- Basement

There are other methods and products available which will achieve the required standard of fire resistance and may be more appropriate for the existing construction in your premises. If there is any doubt about how your building is constructed, then ask for further advice from a competent person.

Fire-resisting floors

The fire resistance of floors will depend on the existing floor construction as well as the type of ceiling finish beneath. If you need to upgrade the fire resistance of your floor it may not be desirable to apply additional fire resistance to the underside of an existing ornate ceiling. In older buildings there may be a requirement to provide fire resistance between beams and joists.

A typical example of a 30-minute fire-resisting timber floor is tongue and groove softwood of not less than 15mm finished thickness on 37mm timber joists, with a ceiling below of one layer of plasterboard to a thickness of 12.5mm with joints taped and filled and backed by supporting timber.

There are other, equally valid, methods and products available for upgrading floors. If you are in any doubt you should seek the advice of a competent person and ensure that the product is installed in accordance with instructions from the manufacturer or supplier.

Fire-resisting glazing

The most common type of fire-resisting glazing is 6mm Georgian wired glazing, which is easily identifiable. Clear fire-resisting glazing is available and can quickly be identified by a mark etched into the glass, usually in the corner of the glazed panel, to confirm its fire-resisting standard. Although this is not compulsory, the marking of glass is supported by the Glass and Glazing Federation, you should check whether the glazing will be marked accordingly before purchase. The glazing should have been installed in accordance with the manufacturer's instructions and to the appropriate standard,[76] to ensure that its fire-resisting properties are maintained.

The performance of glazed systems in terms of fire resistance and external fire exposure should, wherever possible, be confirmed by test evidence. Alternatively, where there is a lack of test information, ask for an assessment of the proposed construction from suitably qualified people.

Fire separation of voids

A common problem encountered with fire separation is fire-resisting partitions which do not extend above false ceilings to true ceiling height. This may result in unseen fire spread and a loss of vital protection to the escape routes. It is important therefore to carefully check all such partitions have been installed correctly.

Breaching fire separation

To ensure effective protection against fire, walls and floors providing fire separation must form a complete barrier, with an equivalent level of fire resistance provided to any openings such as doors, ventilation ducts, pipe passages or refuse chutes.

The passing of services such as heating pipes or electrical cables through fire-resisting partitions leaves gaps through which fire and smoke may spread. This should be rectified by suitable fire stopping and there are many proprietary products available to suit particular types of construction. Such products should be installed by competent contractors.

Decor and surface finishes of walls, ceilings and escape routes

The materials used to line walls and ceilings can contribute significantly to the spread of flame across their surface. Most materials that are used as surface linings will fall into one of three classes of surface spread of flame. The following are common examples of acceptable materials for various situations:

Class 0: Materials suitable for circulation spaces and escape routes

- Such materials include brickwork, blockwork, concrete, ceramic tiles, plaster finishes (including rendering on wood or metal lathes), wood-wool cement slabs and mineral fibre tiles or sheets with cement or resin binding.

Note: Additional finishes to these surfaces may be detrimental to the fire performance of the surface and if there is any doubt about this then consult the manufacturer of the finish.

Class 1: Materials suitable for use in all rooms but not on escape routes

- Such materials include all the Class 0 materials referred to above. Additionally, timber, hardboard, blockboard, particle board, heavy flock wallpapers and thermosetting plastics will be suitable if flame-retardant treated to achieve a Class 1 standard.

Class 3: Materials suitable for use in rooms of less than 4m²

- Such materials include all those referred to in Class 1, including those that have not been flame-retardant treated and certain dense timber or plywood and standard glass-reinforced polyesters.

The equivalent European classification standard will also be acceptable. Further details about internal linings and classifications are available in Approved Document B.[24] Appropriate testing procedures are detailed in BS 476-7[32] and where appropriate BS EN 13501-1.[33]

Further guidance on types of fire-resisting construction has been published by the Building Research Establishment.[34]

B2 Fire-resisting doors

Requirements of a fire-resisting door

Effective fire-resisting doors (see Figure 57) are vital to ensure that the occupants can evacuate to a place of safety. Correctly specified and well-fitted doors will hold back fire and smoke preventing escape routes becoming unusable, as well as preventing the fire spreading from one area to another.

Fire-resisting doors are necessary in any doorway located in a fire-resisting structure. Most internal doors are constructed of timber. These will give some limited protection against fire spread, but only a purpose-built fire-resisting door that has been tested to an approved standard will provide the necessary protection. Metal fire-resisting doors are also available and specific guidance for these follows.

All fire-resisting doors are rated by their performance when tested to an appropriate standard. The level of protection provided by the door is measured, primarily by determining the time taken for a fire to breach the integrity (E), of the door assembly, together with its resistance to the passage of hot gases and flame.

It may be possible to upgrade the fire resistance of existing doors. Further information is available from the Building Research Establishment[77] or the Timber Research and Development Association.[78]

Timber fire-resisting doors require a gap of 2-4mm between the door leaf and the frame. However larger gaps may be necessary to ensure that the door closes flush into its frame when smoke seals are fitted (see BS 4787-1[79] for further information). For fire-resisting purposes the gap is normally protected by installing an intumescent seal in either the door or, preferably, the frame. The intumescent seal expands in the early stages of a fire and enhances the protection given by the door. Additional smoke seals, either incorporated in the intumescent seal or fitted separately, will restrict the spread of smoke at ambient temperatures. Doors fitted with smoke seals have their classification code suffixed with an 'S'.

The principal fire-resisting door categories are:

- E20 fire-resisting door providing 20 minutes fire resistance (or equivalent FD 20S). (Note: Many suppliers no longer provide an E 20 type fire-resisting door.)
- E30 fire-resisting door providing 30 minutes fire resistance (or equivalent FD 30S).
- E60 fire-resisting door providing 60 minutes fire resistance (or equivalent FD 60S).

Timber fire-resisting doors are available that will provide up to 120 minutes fire resistance but their use is limited to more specialised conditions which are beyond the scope of this guidance.

Metal fire-resisting doors

Although the majority of fire-resisting doors are made from timber, metal fire-resisting doors, which meet the appropriate standard, can often be used for the same purpose. The majority of metal fire-resisting door manufacturers will require the use of bespoke frames and hardware for their door sets.

See BS EN 1634-1[35] and BS 476-22[36] for more information.

For detailed guidance refer to Approved Document B.[24]

Glazing in fire-resisting doors

Although glazing provides additional safety in everyday use and can enhance the appearance of fire-resisting doors, it should never reduce the fire resistance of the door. The opening provided in the door for the fire-resisting glazing unit(s), fitted in a proven intumescent glazing system and the fitting of the beading are critical, and should only be entrusted to a competent person. In all cases the door and glazing should be purchased from a reputable supplier who can provide documentary evidence that the door continues to achieve the required rating.

Fire-resisting door furniture

Hinges
To ensure compliance with their rated fire performance, fire-resisting doors must be hung with the correct number, size and quality of hinges. Normally a minimum of three hinges is required, however the manufacturer's instructions should be closely followed. BS EN 1935[37] including Annex B, is the appropriate standard.

Alternative door mountings

Although the most common method of hanging a door is to use single axis hinges, alternative methods are employed where the door is required to be double swing or mounted on pivots for other reasons.

Floor mounted controlled door closing devices are the most common method regularly found with timber, glass and steel doors while transom mounted devices are commonly used with aluminium sections. In each case reference should be made to the fire test report for details as to compliance with the composition of the door assembly including the door mounting conditions.

Self-closing devices

All fire-resisting doors, other than those to locked cupboards and service ducts should be fitted with an appropriately controlled self-closing device that will effectively close the door from any angle. In certain circumstances, concealed, jamb-mounted closing devices may be specified and in these cases should be capable of closing the door from any angle and against any latch fitted to the door; spring hinges are unlikely to be suitable. Further information is given in BS EN 1154.[38]

Rising butt hinges are not suitable for use as a self closing device due to their inability to close and latch the door from any angle

Automatic door hold-open/release devices for self-closing fire doors

These devices are designed to hold open self-closing fire doors or allow them to swing free during normal use. In the event of a fire alarm the device will then release the door automatically, allowing the self-closing mechanism to close the door.

Such devices are particularly useful in situations where self-closing doors on escape routes are used regularly by significant numbers of people, or by people with impaired mobility who may have difficulty in opening the doors.

Typical examples of such devices include:

- electro-magnetic devices fitted to the fire-resisting door which release when the fire detection and warning system operates, allowing a separate self-closer to close the door;

- electro-magnetic devices within the controlled door closing device which function on the operation of the fire detection and warning system ; and

- 'free swing' controlled door closing devices, which operate by allowing the door leaf to work independently of the closing device in normal conditions. An electro-magnetic device within the spring mechanism linked to the fire detection and warning system ensures that the door closes on the operation of the system.

Note: Free swing devices may not be suitable in some situations, such as corridors, where draughts are a problem and the doors are likely to swing uncontrolled, causing possible difficulty or injury to certain people e.g. those with certain disabilities, the elderly and frail, or young children.

Automatic door hold open/release devices fitted to doors protecting escape routes should only be installed in conjunction with an automatic fire detection and warning system incorporating smoke detectors, that is designed to protect the escape routes in the building (see Part 2, Section 2).

In all cases the automatic device should release the fire-resisting door allowing it to close effectively within its frame when any of the following conditions occur:

- the detection of smoke by an automatic detector;

- the actuation of the fire detection and alarm system by manual means e.g. operation of break glass call point;

- any failure of the fire detection and alarm system; or

- any electrical power failure.

Other devices, including self-contained devices which perform a similar function, that are not connected directly to a fire alarm system and are not therefore able to meet the above criteria are available and may be acceptable where a site specific risk assessment can show that they are appropriate. Such devices are unlikely to be suitable for use on doors protecting single stairways or other critical means of escape.

In all cases where a door hold open device is used it should be possible to close the door manually.

A site specific risk assessment should be undertaken before any type of automatic door hold open/release device is installed. If you are unsure about the suitability of such devices in your premises, you should seek the advice of a competent person.

Further guidance about automatic door hold open/release devices is given in BSEN 1155[82] or BS5839-3.[40]

Door co-ordinators
Where pairs of doors with rebated meeting stiles are installed it is critical that the correct closing order is maintained. Door co-ordinators to BS EN 1158[83] should be fitted and fully operational in all cases where the doors are self-closing.

Installation and workmanship
The reliability and performance of correctly specified fire-resisting doors can be undermined by inadequate installation. It is important that installers with the necessary level of skill and knowledge are used. Accreditation schemes for installers of fire-resisting doors are available.

Figure 57: A fire door with smoke seals and infumescent strips

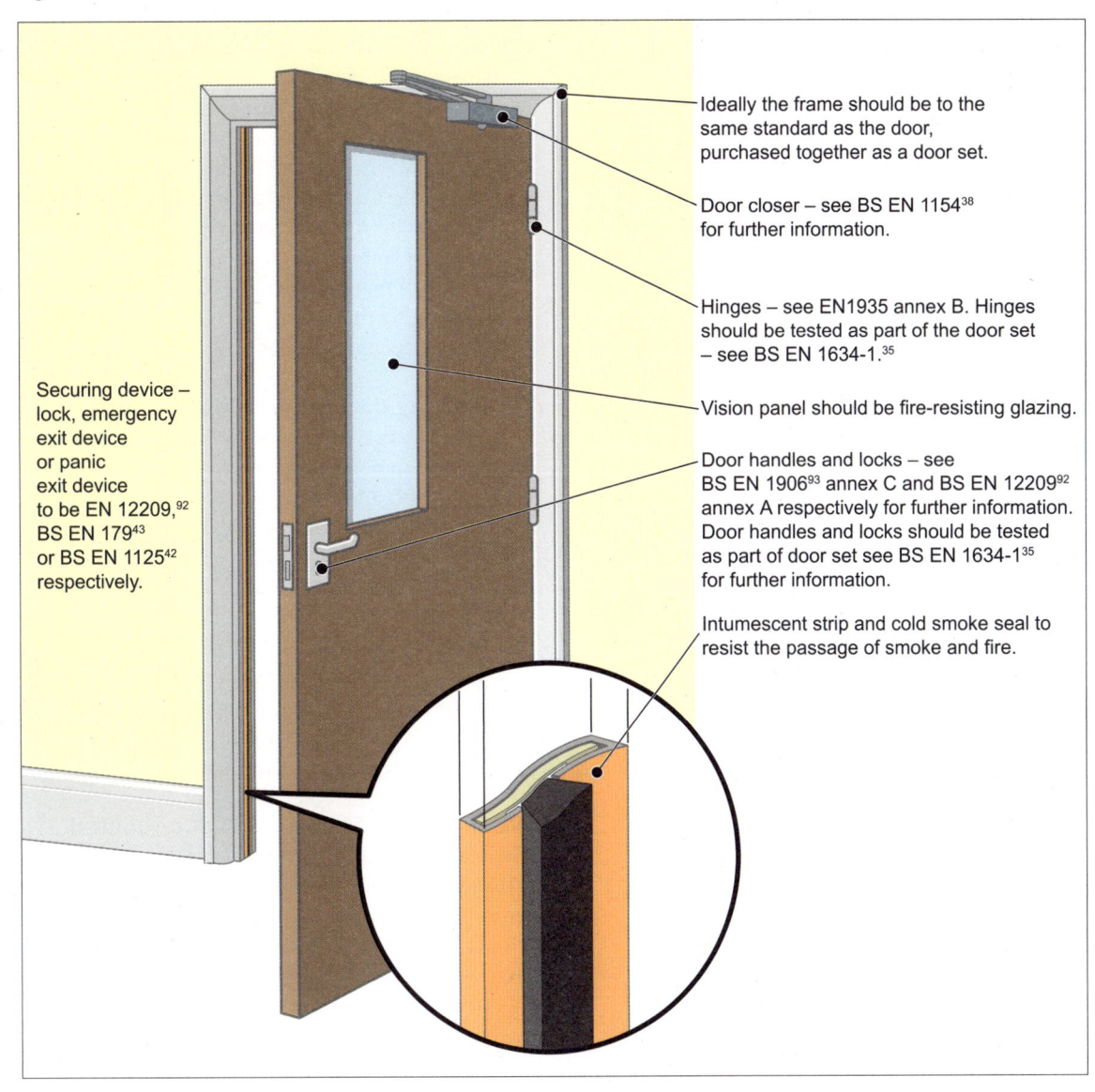

- Ideally the frame should be to the same standard as the door, purchased together as a door set.
- Door closer – see BS EN 1154[38] for further information.
- Hinges – see EN1935 annex B. Hinges should be tested as part of the door set – see BS EN 1634-1.[35]
- Vision panel should be fire-resisting glazing.
- Door handles and locks – see BS EN 1906[93] annex C and BS EN 12209[92] annex A respectively for further information. Door handles and locks should be tested as part of door set see BS EN 1634-1[35] for further information.
- Intumescent strip and cold smoke seal to resist the passage of smoke and fire.
- Securing device – lock, emergency exit device or panic exit device to be EN 12209,[92] BS EN 179[43] or BS EN 1125[42] respectively.

Fire-resisting doors and shutters will require routine maintenance, particularly to power operation and release and closing mechanisms.

Further information is available on fire-resisting doors in BS 8214.[41] If you are unsure about the quality, the effectiveness or the fitting of your fire-resisting doors consult a fire safety expert.

For further guidance on the selection and maintenance of door furniture for timber fire doors refer to the Door Hardware Federation/Guild of Architectural Ironmongers code of practice.[84]

B3 Door-fastening devices

The relationship between the securing of doors against unwanted entry and the ability to escape through them easily in an emergency has often proved problematical. Careful planning and the use of quality materials remain the most effective means of satisfying both of these objectives.

Any device that impedes people making good their escape, either by being unnecessarily complicated to manipulate or not being readily openable, will not be acceptable.

Guidance on fire exits starts from the position that doors on escape routes should not be fitted with any locking devices (electrically operated or otherwise). However, it is accepted that in many cases the need for security will require some form of device that prevents unlimited access but still enables the occupants of a building or area to open the door easily if there is a fire. These devices can take many forms but, in the majority of cases, premises where there are members of the public present or others who are not familiar with the building should use panic exit bar devices (i.e. push bars or touch bars). See BS EN 1125[42] for further information.

Premises that have limited numbers of staff or others who are familiar with the building and where panic is not likely may use alternative devices (i.e. push pads or lever handles). See BS EN 179[43] for further information.

In some larger premises, when only certain staff are on the premises and there is a security issue, it may be acceptable to restrict the number of emergency exits immediately available, e.g. when only security staff are present at night, or prior to opening a premises in the morning. Staff should be made fully aware of any restrictions and the number of exits not immediately available should be limited.

Electrical locking devices

Electrically operated entry control devices have been developed for use as locking devices on fire exits. They fall into two main categories, electromechanical and electromagnetic.

- Electromechanical devices

    Electromechnical devices comprise electromechanical lock keeps and draw bolts, which can be controlled by people inside the premises by entering a code or by using 'smart cards', which have been adapted to control the exit from certain areas. These devices have been fitted in many premises and may be linked to the fire-detection and/or warning system. Experience has shown that these devices can fail to open in a number of ways. They are dependent on a spring mechanism to return the lock keep or draw bolt(s) and are liable to jam when pressure is applied to the door. It is also relatively easy to fit them incorrectly. Electromechanical locking devices are normally unacceptable on escape doors, unless they are fitted with a manual means of overriding the locking mechanism such as a push bar, push pad or lever handle or that they do not rely on a spring mechanism, fail-safe open and are not affected by pressure, in which case the criteria for electromagnetic devices should be applied.

- Electromagnetic devices

    These devices comprise a magnet and a simple fixed retaining plate with no moving parts and are therefore generally considered to be more reliable due to their inherent 'fail-safe unlocked' operation. Electromagnetic locking devices go some way to addressing the particular concerns surrounding electromechanical locking systems. The release of this type of device is controlled by the interruption of electrical current to an electromagnet, either manually via a switch or other means, break-glass point (typically coloured green), or by linking to the fire-warning and/or detection system of the premises.

Time-delay devices on escape routes

A further development is the fitting of a time-delay system to the electronic door-locking device. This delays the actual opening of an exit door for a variable period following operation of the panic bar or other exit device. Periods of between five and 60 seconds can be pre-set at the manufacturing stage or can be adjusted when fitted. These are not usually acceptable for use by members of the public. However, they maybe acceptable for use by staff that are familiar with their operation and are suitably trained in their use.

Management of electronic door-control devices including time delays

The use of such devices may be accepted by enforcing authorities if the responsible person can demonstrate, through a suitable risk assessment for each individual door, both the need and the adequate management controls to ensure that people can escape safely from the premises. In particular:

- Access control should not be confused with exit control. Many devices are available which control the access to the premises but retains the immediate escape facility from the premises.

- In public areas, when push bars are operated on escape doors, they should release the electromagnetic locks immediately and allow the exit doors to open.

- The requirement for exit control should be carefully assessed and should not be seen as a substitute for good management of the employees and occupants.

- All other alternatives should have been explored/evaluated prior to using these devices to ensure they do not affect the safety of occupants.

- The device should be connected to the fire warning and/or detection system.
- The device should incorporate a bypass circuit for immediate release on activation of the fire warning and/or detection system.
- Each door should be fitted with a single securing device.
- The emergency exit doors should be clearly labelled about how to operate them.
- The need for exit control shold be carefully assessed and should not be seen as a substitute for good management of the employees and occupants.
- Adequate control measures should be put in place to ensure the safety of the occupants.

The use of electronic door-locking devices should be considered with particular care in premises with a number of different occupancies. The management of a complicated system of evacuation for many different groups is unlikely to be practicable.

The technical standards in respect of sourcing, maintaining and testing must be extremely high.

When part of the management control system involves trained personnel helping others at these doors, it is vital to ensure these people are available.

The use of exit control devices should not be considered where the number of trained staff is low and members of the public would be expected to operate the devices without help.

In premises where there may be significant numbers of people, the devices should only be considered when linked to a comprehensive automatic fire-detection and warning system (see Section 2 for guidance on appropriate systems). There should be an additional means of manually overriding the locking device at each such exit (typically a green break-glass point).

The use of time-delay systems that prevent the opening of emergency exits for a pre-set time are primarily used to improve security. These add a further layer of complexity to the fire strategy and should not be considered in public areas. They should only be used in non-public areas when all other options such as exterior boundary management have been addressed. Their value in preventing inappropriate use of exits is likely to be transient as the use of the manual override becomes more widely known.

British Standard 8220[44] gives further advice on security in buildings and while this standard does refer to locking devices, it also acknowledges that the balance must remain on the side of emergency escape rather than security.

Appendix C
Historic buildings

General considerations

This appendix offers additional information about listed and historical buildings.

Fire risk assessments conducted for residential care premises which are within a listed or historic building will need to ensure that a balance is struck between ensuring sufficient fire safety measures are in place for the safety of people, yet avoid extensive alterations and helping to maintain the character of the building.

As well as the fire risk assessment it is recommended that a general fire policy statement and manual is compiled. A person must be nominated to take responsibility for all aspects of fire safety. Usually the person charged with the management and control of the premises will be the 'responsible person' under the Order.[1]

The advice and/or consent of a building control body or any other relevant bodies (e.g. English Heritage) should form part of any fire risk assessment that impacts on the character of the building (e.g. replacement of doors, fittings, wooden panelling and decor) or material changes to existing escape routes. An ideal solution is one that is reversible, enabling the historic elements to be reinstated.

A fire safety adviser will be able to suggest alternatives to conventional fire precautions, such as:

- a fire engineering solution;
- upgrading existing doors and partitions in a sympathetic manner to improve their fire resistance; and
- considering the installation of specialist fire-detection or suppression systems.

Should the design and nature of the historic building preclude the introduction of conventional fire safety features, it will be necessary to manage the building in such a way that:

- limits the number of occupants, either staff or members of the public, inside the building;
- limits activities in the building; and
- provides adequate supervision within the building.

In buildings that are open to the public you may wish to designate parts as 'off limits' to the general public. The locking of internal doors or the use of fixed or movable barriers should not restrict alternative escape routes being made available.

Liaison with the fire and rescue service

The responsible person will need to ensure effective liaison with the fire and rescue service to enable them to carry out firefighting operations. These may include information on:

- the provision of water supplies, seasonal ponds, lakes and underground tanks, and any associated pumps;
- difficult access for fire engines;
- particular hazards in the construction features of the building (including asbestos);
- the use of combustible under floor insulation;
- underground vaults ducts and voids where fire may spread unchecked;
- worn stone slabs in stairway construction; and
- the presence of cast iron columns and wrought iron beams.

Emergency planning

An important consideration for the owners and trustees is the protection of valuable artefacts and paintings from the effects of fire. However, the efficient evacuation of all occupants must take precedence over procedures for limiting damage to property and contents. Salvage work should be limited to those parts of the building not directly affected by the fire.

Fire wardens and others tasked with carrying out salvage work should have received formal training, adequate protection and be fully briefed about the health and safety risk assessment carried out to identify the dangers associated with this activity. Further detailed advice on fire safety in historic buildings can be found in the following publications:

- BS 7913: Guide to the principles of the conservation of historic buildings. British Standards Institution

- Heritage under fire: A guide to the protection of historic buildings. Fire Protection Association (for the UK Working Party on Fire Safety in Historic Buildings) 1991, ISBN 0902167944

- The Installation of Sprinkler Systems in Historic Buildings (Historic Scotland Technical Advice Note S). Fire Protection Association (TCRE Division/Scottish Conservation Bureau, Hist.) 1998, ISBN 1 900168 63 4

- Fire Protection Measures in Scottish Historic Buildings: Advice on Measures Required to Minimise the Likelihood of Fire Starting and to Alleviate the Destructive Consequences of Fire in Historic Buildings (Technical Advice Note). TCRE Division/Scottish Conservation Bureau, Hist. 1997, ISBN 1 900168 41 3

- Fire Risk Management in Heritage Buildings (Technical Advice Note). TCRE Division/ Scottish Conservation Bureau, Hist. 2001, ISBN 1 900168 71 5

- Summary and conclusions of the report into fire protection measures for the Royal Palaces by Sir Alan Bailey, following the Windsor Castle fire, 1992.

- The fire at Upton Park. Published by the National Trust.

- Timber panelled doors and fire. Published by English Heritage.

- Fire safety in historic town centres. Published by English Heritage and Cheshire Fire and Rescue Service.

Appendix D
Glossary

These definitions are provided to assist the responsible person in understanding some of the technical terms used in this guide. They are not exhaustive and more precise definitions may be available in other guidance.

Term	Definition
Access room	A room through which the only escape route from an inner room passes.
Accommodation stairway	A stair, additional to that required for means of escape purposes, provided for the convenience of occupants.
Alterations notice	If your premises are considered by the enforcing authority to be high risk, they may issue an alterations notice that requires you to inform them before making any material alterations to your premises.
Alternative escape route	Escape routes sufficiently separated by either direction and space, or by fire-resisting construction to ensure that one is still available irrespective of the location of a fire.
Approved Document B (ADB)[24]	Guidance issued by Government in support of the fire safety aspects of the building regulations.
As low as reasonably practicable	Is a concept where risks should continue to be reduced until you reach a point where the cost and effort to reduce the risk further would be grossly disproportionate to the benefit achieved.
Automatic fire-detection system	A means of automatically detecting the products of a fire and sending a signal to a fire warning system. See 'Fire warning'.
Basement	A storey with a floor which at some point is more than 1,200mm below the highest level of ground adjacent to the outside walls, unless, and for escape purposes only, such area has adequate, independent and separate means of escape.
Child	Anyone who is not over compulsory school age, i.e. before or just after their 16th birthday.
Class 0, 1 or 3 surface spread of flame	Classes of surface spread of flame for materials needed to line the walls and ceilings of escape routes. See Appendix B for further information.
Combustible material	A substance that can be burned.
Compartment wall and/or floor	A fire-resisting wall or floor that separates one fire compartment from another.
Competent person	A person with enough training and experience or knowledge and other qualities to enable them properly to assist in undertaking the preventive and protective measures.

Term	Definition
Dangerous substance	1. A substance which because of its physico-chemical or chemical properties and the way it is used or is present at the workplace creates a risk. 2. A substance subject to the Dangerous Substance and Explosive Atmosphere Regulations 2002 (DSEAR).
Dead end	Area from which escape is possible in one direction only.
Direct distance	The shortest distance from any point within the floor area to the nearest storey exit, or fire-resisting route, ignoring walls, partitions and fixings.
Domestic premises	Premises occupied as a private dwelling, excluding those areas used in common by the occupants of more than one such dwelling.
Emergency escape lighting	Lighting provided to illuminate escape routes that will function if the normal lighting fails.
Enforcing authority	The fire and rescue authority or any other authority specified in Article 25 of the Regulatory Reform (Fire Safety) Order 2005.[1]
Escape route	Route forming that part of the means of escape from any point in a building to a final exit.
Evacuation lift	A lift that may be used for the evacuation of people with disabilities, or others, in a fire.
External escape stair	Stair providing an escape route, external to the building.
Fail-safe	Locking an output device with the application of power and having the device unlock when the power is removed. Also known as fail unlock, reverse action or power locked.
False alarm	A fire signal, usually from a fire warning system, resulting from a cause other than fire.
Final exit	An exit from a building where people can continue to disperse in safety and where they are no longer at danger from fire and/or smoke.
Fire compartment	A building, or part of a building, constructed to prevent the spread of fire to or from another part of the same building or an adjoining building.
Fire door	A door or shutter, together with its frame and furniture, provided for the passage of people, air or goods which, when closed is intended to restrict the passage of fire and/or smoke to a predictable level of performance.
Firefighting lift	A lift, designed to have additional protection, with controls that enable it to be used under the direct control of the fire and rescue service when fighting a fire.
Firefighting shaft	A fire-resisting enclosure containing a firefighting stair, fire mains, firefighting lobbies and if provided, a firefighting lift.
Firefighting stairway	See firefighting shaft.

Term	Definition
Fire resistance	The ability of a component or construction of a building to satisfy, for a stated period of time, some or all of the appropriate criteria of relevant standards. (Generally described as 30 minutes fire-resisting or 60 minutes fire-resisting.) See BS EN 1363-1,[45] BS 476-7[32] and associated standards for further information.
Fire safety manager	A nominated person with responsibility for carrying out day-to-day management of fire safety. (This may or may not be the same as the 'responsible person'.)
Fire safety strategy	A number of planned and co-ordinated arrangements designed to reduce the risk of fire and to ensure the safety of people if there is a fire.
Fire stopping	A seal provided to close an imperfection of fit or design tolerance between elements or components, to restrict the passage of fire and smoke.
Fire-warning system	A means of alerting people to the existence of a fire. (See automatic fire detection system.)
Flammable material	Easily ignited and capable of burning rapidly.
Highly flammable	Generally liquids with a flashpoint of below 21°C. (The Chemicals Hazard Information and Packaging for Supply Regulations 2002[46] (CHIP) give more detailed guidance.)
Hazardous substance	1. See Dangerous substance. 2. A substance subject to the Control of Substances Hazardous to Health Regulations 2002 (COSHH).
Inner room	A room from which escape is possible only by passing through another room (the access room).
Licensed premises	Any premises that require a licence under any statute to undertake trade or conduct business activities.
Material change	An alteration to the premises, process or service which significantly affects the level of risk to people from fire in those premises.
Means of escape	Route(s) provided to ensure safe egress from the premises or other locations to a place of total safety.
Phased evacuation	A system of evacuation in which different parts of the premises are evacuated in a controlled sequence of phases, those parts of the premises expected to be at greatest risk being evacuated first.
Place of reasonable safety	A place within a building or structure where, for a limited period of time, people will have some protection from the effects of fire and smoke. This place, usually a corridor or stairway, will normally have a minimum of 30 minutes fire resistance and allow people to continue their escape to a place of total safety.
Place of total safety	A place, away from the building, in which people are at no immediate danger from the effects of a fire.

Term	Definition
Premises	Any place, such as a building and the immediate land bounded by any enclosure of it, any tent, moveable or temporary structure or any installation or workplace.
Protected lobby	A fire-resisting enclosure providing access to an escape stairway via two sets of fire doors and into which no room opens other than toilets and lifts.
Protected stairway	A stairway which is adequately protected from the rest of the building by fire-resisting construction.
Protected route	An escape route which is adequately protected from the rest of the building by fire-resisting construction.
Refuge	A place of reasonable safety in which a disabled person or others who may need assistance may rest or wait for assistance before reaching a place of total safety. It should lead directly to a fire-resisting escape route.
Responsible person	The person ultimately responsible for fire safety as defined in the Regulatory Reform (Fire Safety) Order 2005.[1]
Relevant persons	Any person lawfully on the premises and any person in the immediate vicinity, but does not include firefighters carrying out firefighting duties.
Self-closing device	A device that is capable of closing the door from any angle and against any latch fitted to the door.
Significant finding	A feature of the premises, from which the fire hazards and persons at risk are identified. The actions you have taken or will take to remove or reduce the chance of a fire occuring or the spread of fire and smoke. The actions people need to take in case of fire. The necessary information, instruction and training needed and how it will be given.
Smoke alarm	Device containing within one housing all the components, except possibly the energy source, for detecting smoke and giving an audible alarm.
Staged fire alarm	A fire warning which can be given in two or more stages for different purposes within a given area (e.g. notifying staff, stand by to evacuate, full evacuation).
Storey exit	A final exit or a doorway giving direct access into a protected stairway, firefighting lobby or external escape route.
Travel distance	The actual distance to be travelled by a person from any point within the floor area to the nearest storey exit or final exit, having regard to the layout of walls, partitions and fixings.
Vision panel	A transparent panel in a wall or door of an inner room enabling the occupant to become aware of a fire in the access area during the early stages.

Term	Definition
Way guidance	Low mounted luminous tracks positioned on escape routes in combination with exit indicators, exit marking and intermediate direction indicators along the route, provided for use when the supply to the normal lighting fails, which do not rely on an electrical supply for their luminous output.
Where necessary	The Order requires that fire precautions (such as firefighting equipment, fire detection and warning, and emergency routes and exits) should be provided (and maintained) 'where necessary'. What this means is that the fire precautions you must provide (and maintain) are those which are needed to reasonably protect relevant persons from risks to them in case of fire. This will be determined by the findings of your risk assessment including the preventative measures you have or will have taken. In practice, it is very unlikely, that a properly conducted fire risk assessment, which takes into account all the matters relevant for the safety of persons in case of fire, will conclude that no fire precautions (including maintenance) are necessary.
Young person	(a) A person aged 16 years, from the date on which he attains that age until and including the 31st August which next follows that date. (b) A person aged 16 years and over who is undertaking a course of full-time education at a school or college which is not advanced education. (c) A person aged 16 years and over who is undertaking approved training that is not provided through a contract of employment. For the purposes of paragraphs (b) and (c) the person: (a) shall have commenced the course of full-time education or approved training before attaining the age of 19 years; and (b) shall not have attained the age of 20 years.

References

The following documents are referenced in this guide. Where dated only this version applies. Where undated, the latest version of the document applies.

1. Regulatory Reform (Fire Safety) Order 2005 SI 2005/1541 The Stationery Office 2005 ISBN 0 11 072945 5
2. Fire Precautions Act 1971 (c 40) The Stationery Office 1971 ISBN 0 10 544071 X
3. Fire Precautions (Workplace) Regulations 1997 SI 1997/1840 The Stationery Office 1997 ISBN 0 11 064738 6
4. Fire Precautions (Workplace) (Amendment) Regulations 1999 SI 1999/1877 The Stationery Office 1999 ISBN 0 11 082882 8
5. Health and Safety (Safety Signs and Signals) Regulations 1996 SI 1996/341 The Stationery Office 1996 ISBN 0 11 054093 X
6. *Safety signs and signals. The Health and Safety (Safety Signs and Signals) Regulations 1996. Guidance on Regulations* L64 HSE Books 1996 ISBN 0 7176 0870 0
7. Dangerous Substances and Explosive Atmospheres Regulations 2002 SI 2002/2776 The Stationery Office 2002 ISBN 0 11 042957 5
8. *Dangerous Substances and Explosive Atmospheres. Dangerous Substances and Explosive Atmospheres Regulations 2002. Approved Code of Practice and guidance* L138 HSE Books 2003 ISBN 0 7176 2203 7
9. *Storage of full and empty LPG cylinders and cartridges. Code of practice 7.* LP Gas Association, 2000.
10. *Maintaining portable electrical equipment in offices and other low-risk environments* Leaflet INDG236 HSE Books 1996 ISBN 0 7176 1272 4
11. Construction (Health, Safety and Welfare) Regulations 1996 SI 1996/1592 The Stationery Office 1996 ISBN 0 11 035904 6
12. *A guide to the Construction (Health, Safety and Welfare) Regulations 1996* Leaflet INDG220 HSE Books 1996 (single copy free or priced packs of 10 ISBN 0 7176 1161 2)

    Health and safety in construction HSG150 (Second edition) HSE Books 2001 ISBN 0 7176 2106 5
13. Disability Discrimination Act 1995 (c 50) The Stationery Office 1995 ISBN 0 10 545095 2
14. BS 8300: 2001 *The design of buildings and their approaches to meet the needs of disabled people. Code of practice* British Standards Institution ISBN 0 580 38438 1
15. ODPM/CACFOA/BFPSA Guidance on reducing false alarms
16. BS 5839-1: *Fire detection and alarm systems for buildings. Code of practice for system design, installation, commissioning and maintenance* British Standards Institution ISBN 0 580 40376 9
17. Manual Handling Operations Regulations 1992 SI 1992/2793 The Stationery Office 1992 ISBN 0 11 025920 3
18. BS 5306-8: *Fire extinguishing installations and equipment on premises. Selection and installation of portable fire extinguishers. Code of practice* British Standards Institution ISBN 0 580 33203 9
19. BS 5306-3: *Fire extinguishing installations and equipment on premises. Code of practice for the inspection and maintenance of portable fire extinguishers* British Standards Institution ISBN 0 5808 42865 6
20. BS 7863: *Recommendations for colour coding to indicate the extinguishing media contained in portable fire extinguishers* British Standards Institution ISBN 0 580 25845 9
21. BS EN 671-3: *Fixed firefighting systems. Hose systems. Maintenance of hose reels with semi-rigid hose and hose systems with lay-flat hose* British Standards Institution ISBN 0 580 34112 7
22. BS EN 12845: *Fixed firefighting systems. Automatic sprinkler systems. Design, installation and maintenance* British Standards Institution ISBN 0 580 44770 7
23. Workplace (Health, Safety and Welfare) Regulations 1992 SI 1992/3004 The Stationery Office 1992 ISBN 0 11 025804 5
24. The Building Regulations 2000: Approved Document B Fire Safety. The Stationery Office. ISBN 0 11753911 2.

25 Local Government (Miscellaneous Provisions) Act 1982 (c 30). The Stationery Office, 1982. ISBN 0 10 543082 X.

26 BS 5395-2: *Stairs, ladders and walkways. Code of practice for the design of industrial type stairs, permanent ladders and walkways* British Standards Institution ISBN 0 580 14706 1.

27 BS EN 50172 (BS 5266-8): *Emergency lighting. Emergency Escape lighting systems* British Standards Institution.

27A BS 5266-1: *Emergency lighting. Code of practice for the emergency lighting of premises.* British Standards Institution.

28 BS EN 1838: *Lighting applications. Emergency lighting* British Standards Institution ISBN 0 580 329925.

29 BS 5499-5: *Graphical symbols and signs. Safety signs, including fire safety signs. Signs with specific safety meanings* British Standards Institution.

30 BS 7974: *Application of fire safety engineering principles to the design of buildings. Code of practice* British Standards Institution ISBN 0 580 38447 0.

31 *Successful health and safety management* HSG65 (Second edition) HSE Books 1997 ISBN 0 7176 1276 7.

32 BS 476- 7: *Fire tests on building materials and structures. Method of test to determine the classification of the surface spread of flame of products.* British Standards Institution.

33 BS EN 13501-1: *Fire classification of construction products and building elements. Classification using test data from reaction to fire tests.* British Standards Institution.

34 *Guidelines for the construction of fire-resisting structural elements.* Building Research Establishment Report BRE 128, 1988.

35 BS EN 1634-1: *Fire resistance tests for door and shutter assemblies. Fire doors and shutters* British Standards Institution ISBN 0 580 32429 X.

36 BS 476-22: *Fire tests on building materials and structures. Methods for determination of the fire resistance of non-loadbearing elements of construction* British Standards Institution. ISBN 0 580 15872 1.

37 BS EN 1935: *Building hardware. Single-axis hinges. Requirements and test methods* British Standards Institution ISBN 0 580 39272 4.

38 BS EN 1154: *Building hardware. Controlled door closing devices. Requirements and test methods* British Standards Institution ISBN 0 580 27476 4.

39 BS EN 1634-2: *Fire resistance tests for door and shutter assemblies. Part 2. Fire door hardware. Building hardware for fire-resisting doorsets and openable windows* British Standards Institution.

40 BS 5839-3: *Fire detection and alarm systems for buildings. Specification for automatic release mechanisms for certain fire protection equipment* British Standards Institution ISBN 0 580 15787 3.

41 BS 8214: *Code of practice for fire door assemblies with non-metallic leaves* British Standards Institution ISBN 0 580 18871 6.

42 BS EN 1125: *Building hardware. Panic exit devices operated by a horizontal bar. Requirements and test methods* British Standards Institution.

43 BS EN 179: *Building hardware. Emergency exit devices operated by a lever handle or push pad. Requirements and test methods* British Standards Institution ISBN 0 580 28863 3.

44 BS 8220: *Guide for security of buildings against crime. Offices and shops* British Standards Institution ISBN 0 580 23692 7.

45 BS EN 1363-1: *Fire resistance tests. General requirements.* British Standards Institution.

46 Chemicals (Hazard Information and Packaging for Supply) Regulations 2002 SI 2002/1689 The Stationery Office 2002 ISBN 0 11 042419 0.

47 The Care Standards Act 2000. HSC 2001/021. Department of Health, 2001.

48 BS 7176: *Specification for resistance to ignition of upholstered furniture for non-domestic seating by composites.* British Standards Institution.

49 *Use of LPG Cylinders, Code of Practice 24. Part 1:1998 – The use of propane in cylinders at residential premises.* LP Gas Association

50 *Use of LPG Cylinders, Code of Practice 24. Part 2: 1998 – The use of butane cylinders at residential premises.* LP Gas Association

51 BS 5287: *Specification for assessment and labelling of textile floor coverings.* British Standards Institution.

52 'Safe use of medical oxygen systems for supply to patients with respiratory disease. European Industrial Gases Association IGC doc 89/02. www.eiga.org

53 The Electricity at Work Regulations 1989. Statutory Instrument 1989 No. 635.

54 The Electrical Equipment (Safety) Regulations 1994. Statutory Instrument 1994 No. 3260.

55 The Construction (Design and Management) Regulations 1994. CONDAM/CDM Regs. HMSO, 1994.

56 *Construction Information Sheet No 51 'Construction Fire Safety'.* Health and Safety Executive

57 *HSG 168 : Fire safety in construction work* (ISBN 0-7176-1332-1). Health and Safety Executive

58 *'Fire prevention on construction sites – The joint code of practice on the protection from fire of construction sites and buildings undergoing renovation'* Fifth edition. London: Fire Protection Association and Construction Federation, 2000 (ISBN 0-90216739-1).

59 *Design, construction, specification and fire management of insulated envelopes for temperature controlled environments.* International Association of Cold Storage Contractors.

60 HTM 87, Textiles and Furniture (Firecode). Department of Health, 1999. ISBN: 0113221394

61 BS 5852: *Methods of test for assessment of the ignitability of upholstered seating by smouldering and flaming ignition.* British Standards Institution.

62 BS 5867-2: *Specification for fabrics for curtains and drapes. Flammability requirements.* British Standards Institution.

63 BS EN 1101: *Textiles and textile products. Burning behaviour. Curtains and drapes. Detailed procedure to determine the ignitability of vertically oriented specimens (small flame)* British Standards Institution.

64 BS EN 1102: *Textiles and textile products. Burning behaviour. Curtains and drapes. Detailed procedure to determine the flame spread of vertically oriented specimens,* British Standards Institution.

65 BS 5866-4: *Blankets suitable for use in the public sector. Specification for flammability performance* British Standards Institution.

67 BS 7175: *Methods of test for the ignitability of bedcovers and pillows by smouldering and flaming ignition sources.* British Standards Institution.

68 BS 5815-3: *Sheets, sheeting, pillowslips, towels, napkins, counterpanes and continental quilt secondary covers suitable for use in the public sector. Specification for counterpanes and continental quilt secondary covers including flammability performance.* British Standards Institution.

69 BS 6807: *Assessment of the ignitability of mattresses, upholstered divans and upholstered bed bases with flaming types of primary and secondary sources of ignition.* British Standards Institution.

70 BS 5722: *Specification for flammability performance of fabrics and fabric combinations used in nightwear garments.* British Standards Institution.

71 Toys (Safety) Regulations, 1995. SI 1995 No 204.

72 Chartered Institution of Building Services Engineers. *CIBSE Guide Volume E: Fire Engineering.* London, CIBSE, 1997.

73 *Design methodologies for smoke and heat exhaust ventilation.* BRE Report 368, Building Research Establishment, 1999.

74 *Effectiveness of sprinklers in residential premises.* BRE Report 204505, Building Research Establishment 2003.

75 BS5588-12:2004 *Fire Precautions in the design, construction and use of buildings – Part 12: Managing fire safety*, British Standards Institution.

76 *A Guide to Best Practice in the Specification and Use of Fire-Resistant Glazed Systems*. Glass and Glazing Federation. May 2005.

77 BRE Information paper IP 8/82: *Increasing the Fire Resistance of Existing Timber Doors*. Building Research Establishment

78 Wood Information Sheet 1-32: *Fire Resisting Doorsets by Upgrading*. Timber Research and Development Association

79 BS 4787-1: *Internal and external wood doorsets, door leaves and frames. Specification for dimensional requirements* British Standards Institution.

80 The Building Regulations 1991: Approved Document M Access to and use of buildings (2004 Edition).

81 BS 5588-8: *Fire precautions in the design, construction and use of buildings. Code of practice for means of escape for disabled people* British Standards Institution.

82 BS EN 1155: *Building hardware. Electrically powered hold-open devices for swing doors. Requirements and test methods*. British Standards Institution.

83 BS EN 1158: *Building hardware. Door coordinator devices. Requirements and test methods*. British Standards Institution.

84 *Code of practice – Hardware for fire and escape doors*. Door Hardware Federation/Guild of Architectural Ironmongers

85 BS EN 3-7: *Portable fire extinguishers. Characteristics, performance requirements and test methods*. British Standards Institution.

86 BS 7177: *Resistance to ignition of mattresses, divans and bed bases*. British Standards Institution.

87 BS 9251: *Sprinkler systems for residential and domestic occupancies code of practice*. British Standards Institution. ISBN 0580 448169.

88 LPC rules for automatic sprinkler installations (incorporating BS EN 12845). The Fire Protection Association.

89 BS 5306-2: *Fire estinguishing installations and equipment on premises specifications for sprinkler installations*. British Standards Institution.

90 BS 5499-4: *Safety signs, including fire safety signs. Code of practice for escape route signing*.

91 BS 5266-6: *Emergency lighting. Code of practice for non-electrical low mounted way guidance systems for emergency use*. Photoluminescent systems.

92 BS EN 12209: *Building hardware. Locks and latches. Mechanically operated locks, latches and locking plates*. British Standards Institution.

93 BS EN 1906: *Building hardware. Lever handles and knob furniture. Requirements and test methods*. British Standards Institution.

94 The Furniture and Furnishings (Fire) (Safety) Regulations 1988 (as amended 1989 and 1993)

Further reading

The latest versions of all documents listed in this section should be used, including any amendments.

Any views expressed in these documents are not necessarily those of the DCLG.

BS EN 1154	Building Hardware. Controlled door closing devices. Requirements and test methods. British Standards Institution.
BS EN 1935	Building hardware. Single-axis hinges. Requirements and test methods. British Standards Institution.
BS EN 1634-1	Fire resistance tests for door and shutter assemblies. Fire doors and shutters. British Standards Institution.
BS 476-22	Fire tests on building materials and structures. Methods for the determination of the fire resistance of non-load bearing elements of construction. British Standards Institution.
BS 4422	Fire. Vocabulary. British Standards Institution.
BS PD 6512-3:	Use of elements of structural fire protection with particular reference to the recommendations given in BS 5588 'Fire precautions in the design and construction of buildings'. Guide to the fire performance of glass. British Standards Institution.
BS 5588-8	Fire precautions in the design, construction and use of buildings. Code of practice for means of escape for disabled people. British Standards Institution.
BS 5588-10	Fire precautions in the design, construction and use of buildings. Code of practice for shopping complexes. British Standards Institution.
BS 5588-11	Fire precautions in the design, construction and use of buildings. Code of practice for shops, offices, industrial, storage and other similar buildings. British Standards Institution.
BS EN 81	Safety rules for the construction and installation of lifts. British Standards Institution.
BS EN 81-70	Safety rules for the construction and installation of lifts. Particular applications for passenger and goods passenger lifts. Accessibility to lifts for persons including persons with disability British Standards Institution.
BS 5839-6	Fire detection and alarm systems for buildings. Code of practice for the design, installation and maintenance of fire detection and fire alarm systems in dwellings British Standards Institution.
BS EN 54-5	Fire detection and fire alarm systems. Heat detectors. Point detectors British Standards Institution.
BS EN 54-7	Fire detection and fire alarm systems. Smoke detectors. Point detectors using scattered light, transmitted light or ionization British Standards Institution.

BS EN 54-11	Fire detection and fire alarm systems. Manual call points. British Standards Institution.
BS 5041-1	Fire hydrant systems equipment. Specification for landing valves for wet risers. British Standards Institution.
BS 5041-2	Fire hydrant systems equipment. Specification for landing valves for dry risers. British Standards Institution.
BS 5041-3	Fire hydrant systems equipment. Specification for inlet breechings for dry riser inlets. British Standards Institution.
BS 5041-4	Fire hydrant systems equipment. Specification for boxes for landing valves for dry risers. British Standards Institution.
BS 5041-5	Fire hydrant systems equipment. Specification for boxes for foam inlets and dry riser inlets. British Standards Institution.
BS 9990	Code of practice for non-automatic firefighting systems in buildings. British Standards Institution.
BS 7944	Type 1 heavy duty fire blankets and type 2 heavy duty heat protective blankets. British Standards Institution.
BS EN 1869	Fire blankets. British Standards Institution.
BS ISO 14520-1	Gaseous fire-extinguishing systems. Physical properties and system design. General requirements. British Standards Institution.
BS 5266-1	Emergency lighting. Code of practice for the emergency lighting of premises other than cinemas and certain other specified premises used for entertainment. British Standards Institution.
BS 5266-2	Emergency lighting. Code of practice for electrical low mounted way guidance systems for emergency use. British Standards Institution.
BS EN 60598-1	Luminaires. General requirements and tests. British Standards Institution.
BS 5499-1	Graphical symbols and signs. Safety signs, including fire safety signs. Specification for geometric shapes, colours and layout. British Standards Institution.
BS EN 1634-1	Fire resistance tests for door and shutter assemblies. Fire doors and shutters. British Standards Institution.
BS EN 1634-3	Fire resistance tests for door and shutter assemblies. Smoke control doors and shutters. British Standards Institution.
BS 8214	Code of practice for fire door assemblies with non-metallic leaves. British Standards Institution.
BS EN 1125	Building hardware. Panic exit devices operated by a horizontal bar. Requirements and test methods. British Standards Institution.
BS EN 179	Building hardware. Emergency exit devices operated by a lever handle or push pad. Requirements and test methods. British Standards Institution.

BS EN 45020	Standardization and related activities. General vocabulary. British Standards Institution.
BS ISO 13784-2	Reaction to fire tests for sandwich panel building systems Part 2: Test method for large rooms. British Standards Institution.
BS 6661	Guide for design, construction and maintenance of single-skin air supported structures. British Standards Institution.
BS 5268-4.2	Structural use of timber. Fire resistance of timber structures. Recommendations for calculating fire resistance of timber stud walls and joisted floor constructions. British Standards Institution.

The design of residential care and nursing homes for older people. HFN 19, Centre for Accessible Environments and NHS Estates.

Northern Ireland Firecode Health Technical Memorandum 84 "Fire Safety in Residential Care Premises".

Fire safety: A guide for care homes in Wales. The fire authorities in Wales and the Care Standards Inspectorate for Wales. 2003.

Design Principles of Fire Safety The Stationery Office 1996 ISBN 0 11 753045 X.

Chemicals (Hazard Information and Packaging for Supply Regulations 2002 SI 2002/1689 The Stationery Office 2002 ISBN 0 11 042419 0and supporting guides: The idiot's guide to CHIP 3: Chemicals (Hazard Information and Packaging for Supply) Regulations 2002 Leaflet INDG350 HSE Books 2002 (single copy free or priced packs of 5 ISBN 0 7176 2333 5), CHIP for everyone HSG228 HSE Books 2002 ISBN 0 7176 2370 X.

Guidance on the Acceptance of Electronic Locks to Doors Required for Means of Escape Published by The Chief and Assistant Chief Fire Officers Association.

Ensuring Best Practice for Passive Fire Protection in Buildings Building Research Establishment 2003 ISBN 1 870409 19 1.

Smoke shafts protecting fire shafts; their performance and design BRE Project Report 79204 Building Research Establishment 2002.

Design, Construction, Specification and Fire Management of Insulated Envelopes for Temperature Controlled Environments International Association for Cold Storage Construction 1999.

Fire safety of PTFE-based material used in building (BRE) Report 274 Building Research Establishment 1994 ISBN 851256538.

Fires and Human Behaviour David Fulton Publishers 2000 ISBN 1853461059.

Management of health and safety at work. Management of Health and Safety at Work Regulations 1999. Approved Code of Practice and guidance L21 (Second edition) HSE Books 2000 ISBN 0 7176 2488 9.

Index

Page numbers in italics refer to information in Figures or Tables.

A

access for firefighters 23, 24, 66, 68, 86, 113, 119, 120, 130
aerosols 13, 19, 46, 48, 50, 112
alarms *see* fire detection and warning systems
alterations 8, 28–9, 42, 46, 52–3, 54, 56, 62, 109, 113, 130, 132, 134
alterations notices 6, 8, 10, 36, 38, 43
arson 12, 16, 18, 19, 46, 48, 57
assembly points 41, 111, 116
automatic fire detection systems 21, 34, 48, 53, 57, 59–60, 62, 81, *82–3*, 86, 101, *103*, 125–6, 128–9, 132

B

basements 47, 49, 84, 99, 101, 104, 132
bedrooms 10
    doors 18, 28
    escape routes 77, 89
    fire detection and warning systems 59, 60
    furniture and furnishings 55
    as refuges 26, 28, 68
building work 8, 46, 52–3

C

candles 12
cavity barriers 73, 85, *123*
ceilings
    fire-resisting 19, 28, 29, 73, 123
    lining materials 13, 19, 55, 124
children 6, 15, 38, 112, 132, 136
    evacuating 24, 26, 111
close down procedures 46, 111, 114, 115
co-operation and co-ordination 7, 37, 38, 39–40, 113–14
competent persons 4, 6, 10, 132
conduction, fire spreading by 16, 17
construction, fire-resisting 28–9, 56, 69–70, 73, 75, 78, 81–5, 88–9, 94, 96, 99, 111, 122–4, 134
contingency plans 112

contractors
    certifying 34, 117
    fire safety training 7, 38, 40, 114
    managing 52–3
    permit to work 18, 53
    risk to 14, 22, 69, 70
    source of ignition 12, 14, 18, 53
convection, fire spreading by 16
cooking 12, 47, 51, 53, 65, 110, 114

D

dangerous substances 7, 13, 14, 40, 46, 112–13, 133
    storage 19, 48–50
day care centres 4, 50
dead ends 73, 80–3, 133
delayed evacuation 26–7, 41, 68, 71, 73, 111
disability *see* people with special needs
display materials 13, 19
doors
    door-fastenings 30, 60, 71, 73, 85, 93, 94, 109, 127–9
    door furniture 125–6, *127*
    final exit 29, 93, 94, 116
    fire-resisting 28, 48, 56, 70, 73, 81, 84, 85, 111, 124–9, 133
    glazing 125
    maintenance 7, 17, 33, 34, 37, 118, 119, 120, 127
    notices 33, 85, 106, 107, 129
    revolving 93
    roller shutter doors 93, 119
    self-closing 15, 29, 65, 73, *75*, 81, 85, 91, 99, 107, 111, 119, 120, 125–6
    sliding 93
    vision panels *127*, 135
    wicket 93

E

electrical safety 12, 16, 18, 46, 48, 51–2, 56, 64

emergency escape lighting 31, 71, 104–5, 111, 133
    tests, checks and maintenance 33, 34, 105, 109, 118, 119, 120

emergency plans 9, 27–30, 36–7, 38–40, 45, 58, 67, 68, 109, 111–12, 113

equipment and machinery 12, 16, 18, 41–2, 46, 47, 48, 50–2, 64, 111

escape routes
    age and construction of the premises 25, 28–9, 46, 53, 68, 69–70
    alternative exits 25, 29, 72, 78, 80, 81, *83,* 84, 86, 88, 95, 96, 132
    basements 47, 49, 84, 101, 104, 132
    corridors 10, 30, 46, 53–4, 70, 71, 84, 85, 86, *87,* 89, 95, 96, 97, *98,* 99, 101–2, 104, 111
    dead ends 70, 80–3, 133
    emergency lighting 31, 104–5, 109, 111, 133
    escape time 25, 27, 28, 70, 71, 73–4, 76
    final exits 29, 30, 64, 76, 86, 88, 93, 94, 95, 133
    hatches 71, 73, 93–4
    inner rooms 79, 80, *81,* 132, 134, 135
    layouts 95–103
    levels of risk 29, 67
    lifts 24, 31, 41, 74, 75, 90, 94, 104, 111, 112, 116, 119, 133
    lobbies 74, 84, 86, *87,* 95, 99, 101, *102,* 133, 135
    maintenance 7, 9, 30–1, 34, 91, 92
    managing 30–1, 128–9
    number required 25, 29, 30, 71, 77, 95
    obstructions 28, 30, 34, 46, 53, 71, 77, 95, 118
    people with special needs 24, 25, 31
    reception areas 90
    roof exits 92, *93*
    signs and notices *30,* 31, 32–3, 94, 104, 106–8, 111
    suitability 30, 68, 71
    tests and checks 33, 34, 35, 118, 119, 120
    training 30, 41, 114
    travel distance 21, 27, 28, 67, 73, 76–7, 79, 80–1, *82–3,* 86, 95, 96, 97, 99–101, 135
    type and number of people using premises 25, 26, 30, 68, 70, 94–5
    widths and capacity of 25, 30, 54, 70–1
    see also doors; stairways

evacuation
    delayed 27, 29, 30, 41, 71, 73, 74–5, 111
    progressive horizontal 27–8, 41, 61, 68, 70, 72–4, 85, 100, 111
    single stage 27, 68, 70–1, 72

extinguishers 22–3, 24, 33, 63–5, 109, 118–9

F

false alarms 22, 60–1, 109, 133

fire
    classes of 63–4
    spread of 16–18, 19, 30, 46, 53, 55–7, 94

fire blankets 47, 65, 109

fire certificates 5, 56

fire dampers 47, 57, 73

fire detection and warning systems 21–2
    automatic fire detection systems 21, 31, 48, 53, 59–60, 81, *82–3,* 86, 95–6, 101, *103,* 126, 129, 132
    auxiliary power supply 22, 62
    false alarms 22, 60–1, 109, 133
    manual call points 21, 59, 60, 61, 111, 118, 126
    monitoring 42, 57, 59, 61, 62, 116
    phased evacuation 21, 25, 61, 134
    quality assurance 117
    record-keeping 35, 62, 109
    smoke detectors 60, 79, 84, 126, 135
    sound levels 22, 28–9, 59
    staff alarms 61
    staged alarms 25
    tests and checks 7, 33, 34, 35, 61–2, 109, 118, 119, 120
    voice alarms *21,* 58, 59, 119

fire drills 40, 41, 42, 109, 115–16

fire marshals 23, 41, 114–15, 116, 131

fire-resisting separation 53, 69–70, 90, 122–4

fire risk assessment
    aims 9
    co-operation and co-ordination 7, 37, 38, 39–40, 113–14
    emergency plans 9, 28, 36–7, 38–40, 45, 58, 67, 68, 109, 111–12, 113, 130
    evaluating risk 9–10, *11*, 15–18, 26, 36, 67, 73
    Fire Safety Order and 5–8, 10, 136
    flexibility 20–1
    historic buildings 5, 73, 130–1
    identifying fire hazards 9, *11*, 12–14, 36, *110, 121*
    identifying people at risk *11*, 14–15, 36, 70, *110, 121*
    information and instruction 7, 9, 32–3, 37, 38–9, 52, 106–8, 112–14
    method 10–11
    plan of action 45, 58
    record-keeping 8, 10, *11*, 36–8, 43, 109, *110*, 111, *121*
    removing or reducing fire hazards 9, 18–21, 36, 45, 46, 57, *110, 121*
    removing or reducing risk to people 5, 20–35, 36, 45, 46, 57, *121*
    reviewing 8, *11*, 42–3, 74, *121*
    sources of fuel 12, 13, 19, 46, 47–50, 57, 67
    sources of ignition 11, 12, 14, 16, 18, 48, 49, 52, 53, 67
    sources of oxygen 12, 13–14, 20
fire safety audit 109, 111
fire safety management 5, 9, 45, 52–3, 71
Fire Safety Order 5–8, 10, 136
fire safety strategy 23, 66, 68, 74, 109, 134
fire stopping 53, 73, 85, 124, 134
fire suppression systems 23–4, 42, 47, 57, 58, 65–6, 70, 76, 81, 85, 109, 111, 112, 115, 119, 130
firefighters' switches 33, 66, 111, 120
firefighting equipment and facilities 22–4
    access for firefighters 23, 24, 66, 71, 86, 113, 118, 119, 120, 130
    extinguishers 22–3, 24, 33, 63–5, 109, 118–9
    fire suppression systems 23–4, 42, 47, 57, 58, 65–6, 70, 76, 81, 85, 109, 111, 112, 115, 119, 130
    firefighters' switches 33, 66, 111, 120

    firefighting lifts and shafts 24, 31, 66, 74, 119, 133
    hose reels 23, 65, 109
    hydrants 66, 118, 119, 120
    maintenance 7, 23–4, 33, 34, 65–6, 109, 118–20
    quality assurance 117
    rising mains 23, 66, 120
    signs 23, 24, 32, 66, 104, 106–7
    tests and checks 33, 34, 35, 65, 118, 119, 120
    training 23, 39, 40–1, 63, 65, 114
fireworks 48
flame-retardant materials 48, 54, 55–6, 124
flammable liquids 13, 19, 48, 49, 51, 57, *63, 64*, 134
floors
    covering materials 13, 19, 56
    fire-resisting 29, 56, 73, 92, 99, 122, 123
foam, fire risk of 13, 56
fuel, sources of 12, 13, 19, 46, 47–50, 57, 67
furniture and furnishings 12, 13, 19, 42, 54, 55–6, 77

G

gas cylinders 14, 20, 41, 49, 50, 53, 112
    see also aerosols; LPG; medical gases
glass, fire-resisting 123

H

hatches 71, 73, 93–4
hazard
    definition 10
    identifying fire hazards 9, *11*, 12–14, 36, *110, 121*
    removing or reducing 9, 18–21, 36, 45, 46, 57, *121*
hazardous materials 19, 41, 42, 50, 134
heating 12, 18, 49, 51, 53, 54
historic buildings 5, 73, 130–1
hose reels 23, 65, 109
housekeeping 41, 46

I

ignition, sources of 11, 12, 14, 16, 18, 48, 49, 52, 53, 67

information and instruction 7, 9, 32–3, 36, 37, 52, 106–8, 112–14

insulated core panels 46, 54

L

ladders 71, 94

laundries 10, 46, 47, 50, 52, 60, 110

lifts
- escape routes 24, 31, 41, 74, 90, 94, 104, 111, 112, 116, 119, 133
- firefighting 24, 31, 66, 74, 119, 133

lighting 12, 16, 31, 105

LPG 13, 49, 51, 53

luminaires 104, 105, 118, 119, 120

M

maintenance
- checklist 118–20
- emergency escape lighting 33, 34, 105, 109, 118, 119, 120
- equipment and machinery 18, 51, 52
- escape routes 7, 9, 30–1, 34, 91, 92
- fire detection and warning systems 7, 33, 34, 35, 61–2, 109, 118, 119, 120
- fire doors 7, 17, 31, 33, 34, 35, 37, 41, 118, 119, 120, 127
- firefighting equipment and facilities 7, 23–4, 33, 34, 65–6, 109, 118–20

medical gases 14, 20, 41, 46, 48, 50, 112
see also gas cylinders

multi-occupied buildings 4, 6
- co-operation and co-ordination 39
- escape routes 86, 92
- fire detection and warning systems 21, 86
- fire risk assessment 15, 86
- information and instruction 107

N

naked flame 12, 18, 50

O

oxygen, sources of 12, 13–14, 20

P

partitions, fire-resisting 53, 81, 122, *123*, 124, 130
see also walls

people at risk
- evaluating 16–18, 72
- identifying *11*, 14–15, 36, 70, *110*, *121*
- removing or reducing risk 5, 20–35, 36, 45, 57, *110*, *121*
- restricting the spread of fire and smoke 53, 55–7
- type and number using premises 25, 26, 30, 42, 68, 71–2, 95

people with special needs
- evacuating 24, 25, 31, 41, 46, 58, 69, 70, 71, 72, 74–5, 111, 115, 116
- fire risk assessment 11, 15, 58
- fire warning systems 22, 59
- information and instruction 39, 105, 106
- risk to 15, 42

personal emergency evacuation plans (PEEPs) 25, 58, 115

piping 14, 20, 41, 53, 56, 124

plans and specifications 36, 37, 95, 100, 111

portable appliance testing (PAT) 51, 56

progressive horizontal evacuation 27–8, 31, 41, 61, 68, *69*, 70, 72–4, 85, 95, 100, 111

protected areas 26, 27–8, 29, 41, 48, 68, 69, 70–4, 76, 97, 100, 111
see also refuge areas

R

radiation, fire spreading by 16, 17
reception areas 90
record-keeping
    emergency escape lighting 105, 109
    fire detection and warning systems 35, 62, 109
    fire risk assessment 8, 10, *11,* 36–8, 43, 109, *110,* 111, *121*
    fire safety training 42, 114, 116
refuge areas 30, 31, 68, 71, 74, 75, 111, 135
    see also protected areas
residential care premises
    emergency plan 38
    escape strategy 26–8
    fire detection and warning systems 60
    fire risk assessment 9, 29, 70
    firefighting equipment and facilities 64
residents
    evacuating 22, 25–8, 41, 58, 68, 70–6, 111, 115–16
    fire detection and warning systems 22
    fire risk assessment 15
    information and instruction 38, 39, 40, 41, 52
    smoking 52
rising mains 23, 66, 120
risk
    definition 10
    evaluating 9–10, *11,* 16–18, 26, 36, 67, 73
    people at risk *11,* 14–15, 16–18, 70
    reducing 5, *11,* 20–35, 45, 46, 57
roll calls 113, 116
roller shutter doors 93, 119
roofs
    fire risk assessment 57, 60
    roof exits 92, *93*

S

security 29, 34, 57, 94–5, 127, 128
shafts, firefighting *see* lifts
signs and notices 23, 24, *30,* 31, 32–3, 51, 66, 104, 106–8, 129
    information and instruction 39, 52, 85, 94
single stage evacuation 27, 68, 70–1, 72
smoke
    control of 33, 47, 55, 57, 60, 66, 73, 85, 94, 109, 120, 124–5
    dangers of 17, 55
    smoke dampers 47, 57, 73
    spread of 16–18, 30, 46, 53, 55, 94
Smoke and Heat Exhaust Ventilation Systems (SHEVS) 57, 60, 94, 109, 120
smoke detectors 60, 79, 84, 126, 135
smoking 12, 16, 18, 46, 50, 52, 114
sprinklers 23–4, 42, 47, 57, 65–6, 70, 111, 112, 115, 119
staff
    evacuating residents 25–8
    fire drills 40, 41, 42, 109, 115–16
    fire safety training 7, 9, 21, 22, 23, 26, 29, 36, 39, 40–2, 63, 65, 114–16
    information and instruction 7, 9, 32–3, 38–9, 107–8, 112–14
    number of 68, 70–1, 72, 75, 129
    staff alarms 61
    see also contractors
stairways
    accommodation stairways 48, 86, 90, 132
    assistance on 75
    basements 84, 99, 101
    bypass routes 89, *90*
    external 91, 133
    hazards in 10, 30, 46, 48, 53–4
    protected 70, 71, 73, 76, 86–9, 91, 95, 97, 99, 101, *102,* 135
    spiral and helical 92
    width and capacity of 73
storage 19, 42, 46, 47–50, 51, 57

T

telephones 32, 33, 66, 113

training
    fire safety 7, 9, 21, 22, 26, 30, 36, 38, 39, 40–2, 114–16
    firefighting equipment 23, 39, 41, 63, 65, 114, 115
    record-keeping 42, 114, 116

V

vandalism 12, 19, 64, 65

ventilation systems 13–14, 17, 20, 42, 49, 51, 57, 60, 120

vision panels 79, *127,* 135

voids 48, 53, 60, 67, 73, 123, 130

W

walls
    fire-resisting 28, 29, 48, 55, 56, 73, 75, 78, *82–3,* 91, 122
    insulated core panels 46, 54
    lining materials 13, 19, 55, 124
    see also partitions

waste and packaging, combustible 13, 16, 19, 46, 52, 57

way guidance systems 31, 105, 136

windows, fire-resisting 91, 123
    vision panels 79, *127,* 135

Part 2 • Further guidance on fire risk assessment and fire precautions